Thomas M Cleemann

The Railroad Engineer's Practice

Thomas M Cleemann

The Railroad Engineer's Practice

ISBN/EAN: 9783337415730

Printed in Europe, USA, Canada, Australia, Japan

Cover: Foto ©berggeist007 / pixelio.de

More available books at **www.hansebooks.com**

THE
RAILROAD ENGINEER'S PRACTICE,

BEING

A SHORT BUT COMPLETE DESCRIPTION OF THE DUTIES OF THE YOUNG ENGINEER IN PRELIMINARY AND LOCATION SURVEYS AND IN CONSTRUCTION.

Second Edition. Revised and Enlarged.

BY

THOMAS M. CLEEMANN, A. M., C. E.

NEW YORK:
THE ENGINEERING NEWS PUBLISHING CO.,
1883.

COPYRIGHT:
THOMAS M. CLEEMANN,
1882.

ATKIN & PROUT, PRINTERS, NO. 12 BARCLAY STREET,
New York.

TO
W. H. WILSON, Esq.,

the Chief Engineer under whom the writer began the practice of his profession, on the Pennsylvania Railroad, and whose uniform kindness and interest in his welfare have been continual causes of pleasure and gratitude, this book is respectfully dedicated by   T. M. C.

PREFACE.

In the first edition of this book many typographical errors occurred, which were a source of great mortification to the author, as they were, of course, due to a want of sufficient care on his part in the proof-reading. He was not before aware of the great difficulty of insuring absolute accuracy. Now, however, he believes he has secured a perfect text, not only in the old portions of the book, but likewise in the numerous additions that have been made. He begs to thank those engineers who have, by their demand for the first edition, caused this second one to appear. Several have made valuable suggestions to him, either in regard to what had been previously left out, or in the simplification of formulæ, which are gratefully acknowledged.

TABLE OF CONTENTS.

	PAGE.
PRELIMINARY SURVEY.	
Inspection by Chief Engineer	1
Method Pursued by Principal Assistant Engineers	1
Organization of Parties	1
Starting a Survey	1
Duties of Transitmen, Leveller and Topographer	2
Convenient Form of Clinometer	4
Form of Transit-Book	4
Topographer's Duties	5
The Paper Location	6
Form for Excavation and Embankment	7
Slopes of Cuts and Fills	7
Levelling by Barometer	8
LOCATION.	
Organization of Party	8
Starting the Location	8
Problems in Curves	9
Transit Points	19
Form of Field-Book	19
Location in a Mountainous Country	19
Location in an Undulating Country	20
Minimum Radius of Curvature	20
Maximum Grade	20
Equation of Grades and Curves	21
Grade at Foot of Mountain Inclines	22
Tangent Between Reversed Curves	22
Actual Smallest Radius of Curvature	22
Division of Line into Sections	23
Letting the Contracts	23
Camp Equipage	23
Adjustments of the Transit	25
Adjustments of the Level	26

CONSTRUCTION.

	PAGE.
Principal Assistant Engineer; his Party and Duties	27
Retracing Line	27
Guarding "Plugs"	28
Form of Note-Book	29
Setting Slope-Stakes	30
Form of Field-Book	31
Calculating Cross-Section Areas	32
Calculating Cubic Contents	32
Slope Ditch	35
Drains for Wet Slopes	35
Estimates of Work	35

CULVERTS.

Finding Water-Way	36

BOX CULVERTS.

How to Lay Out	36
Proper Size	37

OPEN CULVERTS — 37

CATTLE-GUARDS — 37

OPEN PASSAGE-WAYS — 38

STONE ARCHES.

Formulæ	40
Centring	43

RETAINING WALLS.

Formulæ	44

TUNNELS.

The "Heading"	47
A Summit	47
Shafts	47
Arching	47
Form of Intrados	48
Timbering	49
Size of Excavation	49
Blasts in Excavation	50
Dimensions of Tunnels	51
Bond in Tunnels	51

BRIDGES.

	PAGE.
Formulæ	52
Greatest Variable Load	60
Weight of Bridge	61
English and American Practice	62
Proper Place for Pin	62
Howe Truss "Keys"	63
Howe Truss Upper Chord	63
Factor of Safety	64
Wöhler's Law	65
Howe Truss Splice	66
Sizes of Timber	66
Notches for Angle-Blocks	67
Washers for Rods	67
Cast-Iron Tubes	67
Dowel Pins	68
Bracing	68
Wind Pressure	68
Erection of Bridges	69
Floor System	69
Camber	70
Economical Height	71
Rivetting	71
Pin Connections	73
Wrought-Iron Upper Chords	75
Wrought-Iron Columns	75
Least Radius of Gyration by Calculation	76
Least Radius of Gyration by Experiment	76
Sections of Columns	77
Specification for Wrought Iron	79
Trestle-Work; Wood	82
Trestle-Work; Iron	83
Measurement of Bridge Spans	88
Triangulation	94

MASONRY.

Contractors' Tricks	96
Rankine's Rule	96
Preparation of Mortar	97
Cement Mixing and Using	97

FOUNDATIONS

Crushing Strains of Stone	97
Loads per Square Foot	97
Experiments of Sir Charles Fox and Mr. Leonard	98
Rip-Rapping	98
On Gravel in Water	98

PILE-DRIVING.

Formulæ	99
Safe Load on Piles	101
Proper Diameter	101
Bearing Power of Discs	102
Through Boulders and Gravel	102
Through Sand	102
Surface Friction of Cast-Iron Cylinders	102
Water Jet for Driving Piles	103
Bracing of Piles	103

TRACK-LAYING.

Re-running Centre Line	104
Rule for Track-Laying	104
Method of Work	104
Bending Rails	104
Elevation of Outer Rail on Curves	105
Curves of Adjustment	105
Widening of Gauge on Curves	108
Cross-Sections of Road-Bed	109
Specifications of Road-Bed	109
Rail Joints	112

SWITCHES.

Formulæ	113
Frogs, Turnouts, etc.	113

CROSS-TIES.

How Made and Piled	117
Public Road-Crossings	117

RAILS.

Manufacture of Iron Rails	117
Specifications for Steel Rails	118
Composition of Steel Rails	120
Tests of Rails	121

WATER STATIONS. PAGE.
   Use of Water in Engines - - - - - 121
   Gravity Supply - - - - - - 122
   Stand-Pipe - - - - - - - 123
   Tanks - - - - - - - 123
   Reservoirs - - - - - - - 123
   Steam and Wind Power - - - - 123
COALING STATIONS - - - - - - 124
PASSENGER STATIONS.
   Description of Cresson Station - - - 124
TELEGRAPH LINE - - - - - - 125
APPENDIX.
   Specification for Construction of Road-Bed - - 127
   Economical Height of Bridge - - - - 135

PRELIMINARY SURVEY.

The Chief Engineer, from an inspection of the various maps of the country he can obtain, and a personal examination of the ground, decides where it will be necessary to run lines to determine which is the cheapest that can be built, having a due regard to the subsequent cost of operation and maintenance, and gives the necessary orders for such lines to the Principal Assistant Engineers.

The method pursued by the Principal Assistant Engineer differs according to the character of the country, and the time that can be devoted to the preliminary survey. A quick, rough method of gaining the requisite information for the location will first be given, and afterward, one more exact—that pursued on the Bennett's Branch Extension of the Allegheny Valley Railroad. The latter is especially recommended where the means of the company will admit of the more accurate work, and where it may not be a matter of policy to begin the construction of the road at the earliest possible moment.

Each Principal Assistant organizes a party which consists as follows: Principal Assistant, Transitman, Leveller, Topographer, Level Rodman, Slope Rodman, Flagman, two Chainmen, three or more Axemen. An Axeman provides a number of stakes in advance and numbers them consecutively from 0, shaving off a smooth place for that purpose, and drives the first one—usually driven flush with the ground, and called a "plug"—at the place indicated by the Principal Assistant. The latter then starts ahead with the Flagman, and the Transitman sets his transit over the first stake or plug. The Principal Assistant, having decided where he wishes to run the line, sets up the flag. The Chainmen instantly begin chaining toward it, the hind one "lining" the head one, and an Axeman driving the stakes, one every 100 feet, in the order of marking. The Transitman takes his sight, reads only the needle to quarter degrees, records the reading, and starts off for the flag. On arriving there, he sets up ready for another sight, which the Principal Assistant is ready to give him, by waving his handkerchief if the Flagman has not had time to come up. In an open country, the speed of the Chainmen should govern the speed of the party. When there is much underbrush, the Principal Assistant may require several Axemen to clear the way.

The Leveller follows the Transitman as closely as possible, taking levels on every stake, and, if necessary, on abrupt intermediate changes of the slope. His Axeman makes "pegs" (or turning points) and cuts down brush obstructing his view.

The Topographer follows a day behind the Transitman and Leveller. He is provided with a thin box, with a hinged cover on the end, which serves both as a portfolio and a drawing-board. There should be some oiled cloth fastened to it for keeping the paper dry. The paper is tacked to the board with thumb tacks; a convenient size of sheet is 21 × 16 inches. The Topographer has obtained at night from the Transitman and Leveller their notes of the day, and plotted the line on a scale of 400 feet to the inch, noting the elevations at the stations. He takes this into the field with him. His Slope Rodman and Axeman go ahead and measure the transverse slopes, laying a rod upon the ground at each station, and upon it a clinometer;

with a tape they measure the distance to where the slope changes, and then measure the new slope and its length, and so on. This is done on each side of the line, and is noted in the Rodman's book in one of two ways: the direction of the slope being indicated either by the signs + and —, or by the inclination of the line dividing the numerator from the denominator of a fraction in which the numerator is the angle, and the denominator is the distance. The notes are given to the Topographer, and with the help of the elevations already obtained from the Leveller, he sketches in the contours. To facilitate this, he uses a table which gives the horizontal distance between two contours taken ten feet apart for each degree, as follows:

 1° is 573 per 10 feet rise.
 2° " 286 "   "   "   "
 3° " 191 "   "   "   "
 4° " 143 "   "   "   "
 5° " 114 "   "   "   "
 6° "   95 "   "   "   "
 7° "   81 "   "   "   "
 8° "   71 "   "   "   "
 9° "   63 "   "   "   "
10° "   57 "   "   "   "
11° "   51 "   "   "   "
12° "   47 "   "   "   "
13° "   43 "   "   "   "

It is well for him to commit this table to memory.

He estimates distances beyond those measured by the Rodman, and so puts in distant hills, &c. For this it is convenient to have a pocket sextant. The Rodman runs out to different distances, according to the nature of the ground. If the country is level, he may run out 500 feet on each side of the centre-line, only doing so, perhaps, at intervals of 500 feet, or it may be necessary to run out that distance at every station. If the country is hilly, it may be sufficient to run out only 100 feet at each station.

A convenient form of clinometer is formed of a square board, with a string and bullet:

The more exact method is thus described in a private letter written by Mr. A. B. Nichols, in 1870, when he was Principal Assistant on the Bennett's Branch Extension of the Allegheny Valley Railroad:

"I always run experimental with the *vernier* as follows: Going ahead by myself, I select about the spot where I want to 'plug,' and let the Transitman take a sight on me, setting his vernier to the nearest quarter degree (except in special cases). I have the head Chainman carry a sight-staff, and set all the stakes with the transit. The head Chainman then sets the fore-sight plug when he arrives at the end of the sight. I use the needle merely as a check on the vernier. I think it better to set the

FORM OF TRANSIT NOTE BOOK.

'H.'

Sta.	Angle.	Deduced Course.	Needle.	Remarks.
8	....	....	....	+ 40 and + 73 edges of turnpike.
7	....	....	....	
6	....	....	....	+ 20 and + 55 stream.
5	L. 6°	N. 32¾° W.	N. 32¾° W.	
4				
3				
+ 30	R. 13° 25′	N. 26¾° W.	N. 26¾° W.	
2	....	....		
1	....	....		
0	....	....	N. 40° W.	

stakes with the transit, as they are more reliable as references as location, and in an open country they can be set as fast as the Leveller can run (beyond which speed there is no use in running), while in a wooded section there is plenty of time to set them while the Axemen are clearing. In thick woods, the Principal Assistant's voice has often to be taken as the guide ahead. The bench marks should be marked with the number of the station immediately preceding, and the distinctive letter of the line. Thus, if there happen to be a bench at 7 + 60 of 'II' line, it should be marked B. M. | 7 'II.'

"In regard to the Topographer's duties, I do not like the system of putting in the topography in the field. It has always been the custom, I believe, to run experimental one day and locate over the next. Mr. J. A. Wilson's method differs somewhat from this, and, I think, with reason. Topography put in in the haste that is inevitable in the field, is liable to many errors, and locations made on the previous day's experimental may not suit the country ahead. Mr. Wilson's method is to run all the necessary lines, take all the necessary notes, and then go into office quarters and work the maps up, and make a paper location which may then be run in and modified in the field.

"In 'Morrison's Cove' Mr. Linton took charge of the topographical department, taking the topography notes himself. His instruments were: a pocket compass, mounted on a light tripod, a Locke's level, and a small slope-board. His method of proceeding was as follows:

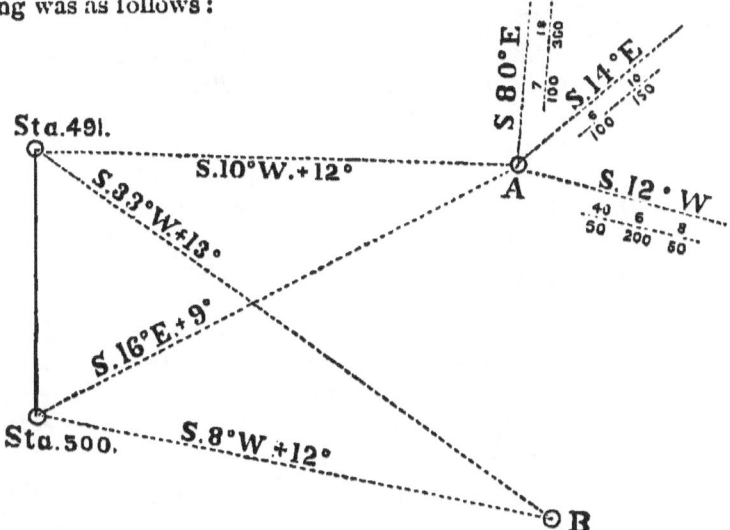

"Say that A is a tree on a hill, and B another point on another hill. He would set his compass up at 491, for instance, take the courses to A and B, and measure the vertical angle with his slope-board. He would then proceed to A and take slopes in all directions, and in like manner from B, using his slope-board and level for heights and slopes. Then going to another station, as 500, he would fix the points A and B by other courses and slopes. Hollows can be shown by running a course up and taking slopes to right and left. By that means he could show the topography sometimes a half a mile from the line. I have known his elevations, say at A, to come within a foot of each other at the distance of half mile from the line, deduced from vertical angles taken with the slope-board, as from 491 and 500; seldom over two feet difference.

"The slope-board is a modification of the square-board and bullet. It will read to quarters of a degree, is furnished with sights,* and is used as follows:

"The Assistant Topographer takes his stand at the station, and gives the right angle to the line by means of a right-angle box, or otherwise. The Slope Rodman measures out the horizontal distance with a ten-feet-long pole to change of slope, and sights back on the man at the station, taking a point on the other's person (previously determined) at the same height above ground as his own eyes. He reads the slope, calls it and the distance out, and in the meanwhile the man at the station, be it the Assistant or the other Rodman, checks the slope by sighting on his person. Rodman No. 1 then measures ahead to the next change, while Rodman No. 2 comes up to *change* No. 1; they measure the slope, and so on. The Assistant keeps the books, and should be furnished with a 'Jacob's staff' and compass for taking buildings, and while so engaged the Rodmen can measure the sizes of said building with a tape, or can go on taking slopes, which they afterward report to the Assistant. Slopes should never be estimated, except one at the end of a series, and then it should be so marked, and the contours derived from it should be dotted on the map to avoid errors in location. In taking short slopes, one Rodman can take the right and the other the left of the line, thus facilitating matters."

From an inspection of the maps, it will be seen on which routes it is necessary to have paper locations made. A

---

* Mr. Linton's improved **slope instrument** may now be obtained at **mathematical instrument makers.**

paper location is such a line drawn upon the plan as may appear, taken in connection with the profile, to require the least excavation and embankment. The following is an excellent method of obtaining the cheapest location on the preliminary map: Having located a trial line by inspection, a profile is made, and grades assumed and drawn. A horizontal plane is supposed to pass through the point on the grade line at each station, and a point, in its line of intersection with the ground surface opposite the station, is marked in red. Having plotted these red points for a sufficient distance, they are connected by a line, which will resemble a contour line, and actually becomes one when the grade is level. The nearer the paper location can be drawn to this line, the less will be the excavation and embankment. If it coincides with it, the line will be a surface line.

Having made the locations on such lines as are considered desirable, a new profile is made from an inspection of where the located line cuts the contours, and cross-sections are plotted on a scale of ten feet to the inch, or on Trautwine's cross-section paper. From these cross-sections, the amounts of excavation and embankment are calculated, and the results embodied in a table of the following form:

Sta.	Distance.	Eleva.	Grade.	Cut.	Fill.	Areas.			Solids.		
						Excavation.		Embankment.	Rock.	Earth.	Emb.
						Rock.	Earth.				

The slopes of the cuts are sometimes assumed as follows:

When the slope of the ground is 20° or less, make the slope 1½ to 1.

When the slope of the ground is 20° to 35°, make the slope 1 to 1.

When the slope of the ground is over 35°, make a vertical wall. Rock stands at ¼ to 1.

Embankment is generally taken as sloping 1½ to 1.

From the calculated amount of excavation and excess of embankment, an estimate is made of the costs of different routes, and it is thus found which lines it will be necessary to actually locate in the field, in order to obtain a closer estimate, or for the purpose of constructing.

In crossing a mountainous country, the Chief Engineer often decides which of several passes it may be necessary to have lines run through, by observing their heights with an Aneroid barometer. Those, of course, are thrown out of the number requiring a more exact determination, which have a higher summit, without any compensating advantages in the way of the cost of the work. The Aneroid is first read at some point of known elevation, and then taken to the summit whose elevation is desired; the difference of elevation in feet is given by the following formula :

$$60664 \left(1 + \frac{T + t - 64}{901}\right) \log \frac{H}{h}$$

in which H and h are the heights in inches of the barometer, at the lower and upper stations, and T and t are the temperatures of the air in degrees Fahrenheit at the times of observation at the same stations.

LOCATION.

The locating party is organized somewhat differently from the preliminary one. We have: Principal Assistant, Transitman, Leveller Level-Rodman, Front Flagman, Back Flagman, two Chainmen, two or more Axemen.

The axeman who drives the stakes now carries tacks,

or better, lath nails, with him, and drives one in the plug at the point where the Transitman is to set up. The Transitman uses the vernier entirely, not using the needle, unless as a check on the tangents. The curves are all run on the ground, and the stakes which come upon them set with the transit. The Leveller keeps close up to the Transitman, and constantly reports the heights to the Principal Assistant. The tangents are generally fixed by the paper locations, and the usual object is to run a given curve from one end of a tangent, and strike the hill with the point of tangent ($P.\ T.$) at a given elevation, viz., that on the paper location. Other problems will also often arise. The following are the most useful:

Problem 1. To change a curve so that it shall come out in a parallel tangent at a given distance from the old tangent, by changing the radius. (From "Haslett & Hackley's Pocket-Book.")

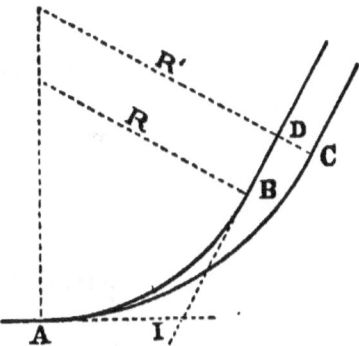

To change the curve $A\ B$ so that it shall come out at C,

$$R' = R \pm \frac{DC}{1 - \cos.\ I;}$$

or otherwise;

Degree of curve $A\ C$ = degree of curve $A\ B \mp \dfrac{8\ D\ C}{7\ n^2}$

in which n is the number of 100-feet chords in $A\ B$.

Problem 2. To change the origin of a curve so that it may pass through a given point.

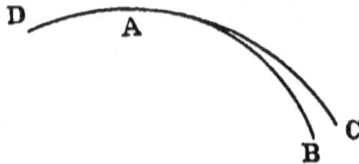

To move A, the point of compound curvature, so that the curve $A B$ will pass through the point C.

Take the distance $B C$, divide it by $A B$, and multiply by 57.3, and we get the difference in deflection $C A B$, which, divided by the number of stations in $A B$, gives the difference in deflection per station (or look in the table of natural sines for $\dfrac{B C}{A B}$, from which is obtained the angle $C A B$, which is to be divided by the number of stations). Then take the difference between the degrees of the curves $A B$ and $D A$; then, the difference between the degrees of the curves is to the difference in deflection per station as $A B$ is to the number of feet forward or backward we must go, on the curve $A D$, to strike the point C.

This is expressed by formula as follows:

$$x = \frac{R\, r\, d}{l\,(R - r)}$$

in which R and r are the radii of the curves, l is the length of the last curve and d is the distance the point of tangent is to be moved over; x being the distance the point of compound curvature is to be moved.

Problem 3. Having located a compound curve terminating in a tangent, it is required to change the point of compound curvature so that the curve will terminate in a tangent parallel to the located tangent, at any required distance perpendicular thereto. Divide the required distance between parallel tangents by the difference of radii of the two last branches of the curve. From the cosine of total amount of curvature in the last branch subtract or add this quotient. The remainder, or sum, will be the natural cosine of the amount of curvature required for the last radius.

This may be expressed by a formula as follows:

$$\cos.(I - x) = \cos. I \pm \frac{d}{R - r}$$

in which I is the total angle in the second branch of the curve, R and r are the radii, d is the distance the tangent is to be moved over, and x is the angle by which the total angle in the last branch is to be increased or diminished (the total angle of the first branch being equally diminished or increased). If the radius of the first curve is larger than that of the second, the upper sign is to be used for moving the curve forward, and the lower one backward; if the radius of the second curve is longer than that of the first, the upper sign will move it backward, and the lower one forward.

Problem 4. Having located a compound curve terminating in a given tangent, it is required to change the point of compound curvature, and also the length of the last radius, so as to pass through the same terminating point with a given difference in the direction of the tangent.

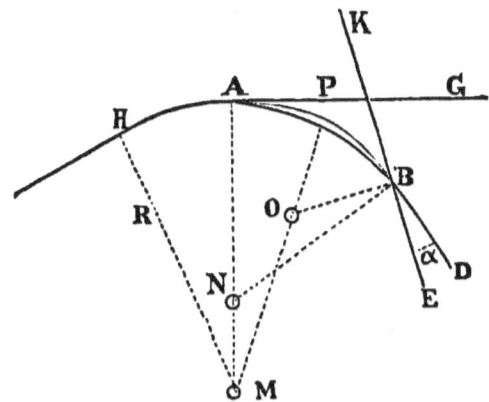

Having the curve HA and the curve AB, with tangent BD, it is required to continue the first curve from A to such a point, P, that the tangent at B will have the direction BE.

Continue the curve HA to the point P, given by the following equation:

$$\text{Cotan. } \tfrac{1}{2} A M P = \frac{R}{R'} (\text{cotan. } \tfrac{1}{2} A N B + \text{cotan. } \tfrac{1}{2} \alpha) - \text{cotan } \tfrac{1}{2} \alpha.$$

The curve from P to B is, of course, found by measuring the total deflection angle, and dividing by the number of stations.

We can, by means of this problem, connect two curves running toward each other with a third one. Let A and B be points on the respective curves. We wish to continue the curve HA past A to some point, P, from which to run some third curve connecting with the other curve at B (the tangent BE being common to the last two curves).

Measure the angle $G A B = \tfrac{1}{2} A N B$; also the angle $K B A = \tfrac{1}{2} A N B + \alpha$; and the distance $A B = 2 R' \sin. \tfrac{1}{2} A N B$. Then calculate $A M P$ from the above equation; dividing by the degree of the curve $H A$ gives the distance $A P$ in stations. From $P O B = A N B + \alpha - A M P$ and the distance $P B$, we can find the degree of the curve PB.

Problem 5. To change the radius of a curve so that it will come out in a given tangent.

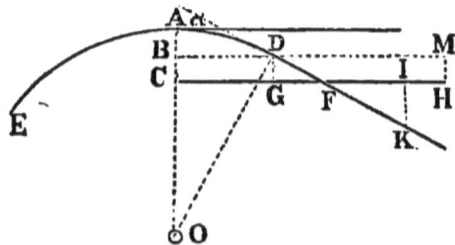

To change the radius of the curve $E D$ so that it will come out in the tangent $C H$.

Having run the curve until the tangent $D K$ is nearly parallel to $C H$, measure the offsets $D G$ and $I K$, and the distance $G I$. Calculate $G F$ and then $\alpha \ (= \tan.^{-1} \frac{D G}{G F})$. We could also measure this angle directly by measuring off $M H = D G$ and taking a sight on M. Then $A C = A B + B C = R \text{ ver. sin. } \alpha + D G$.

We then find the new radius by Problem 1.

$$R = R \pm \frac{A\,C}{ver.\ sin.\ I}$$

(I = the former total angle minus α).

Problem 6. Having located a curve connecting two tangents, it is required to move the middle of the curve any given distance either toward or from the vertex.

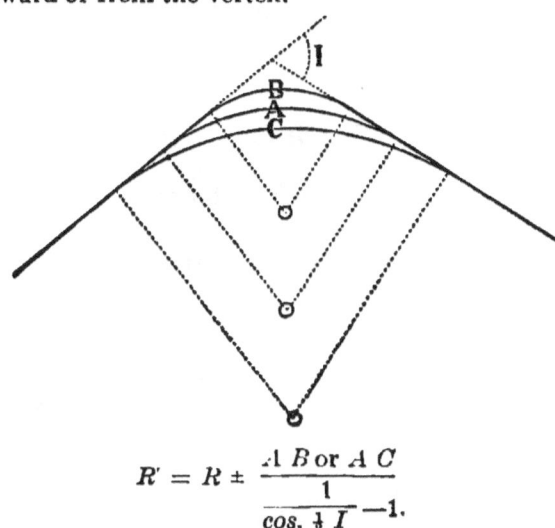

$$R' = R \pm \frac{A\,B \text{ or } A\,C}{\dfrac{1}{\cos.\ \frac{1}{2} I} - 1}$$

Problem 7. To change the origin of a curve so that it shall terminate in a tangent parallel to a given tangent at a given distance from it.

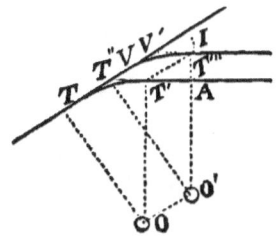

Let $T\,T$ be the curve, $V\,A$ the given tangent, and $V'\,T'''$ the parallel tangent.

$$\text{Then } T\,T' = \frac{A\,T''}{sin.\ I}$$

Problem 8. To find how far back it is necessary to go from the point B, to strike the point C with a curve of given radius; BC being known.

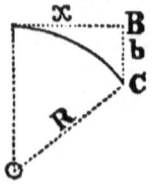

$$x = \sqrt{b(2R - b)} = \sqrt{bD}$$

approximately, where D is the diameter of the curve.

Problem 9. To draw a tangent to a curve from a point outside of it.

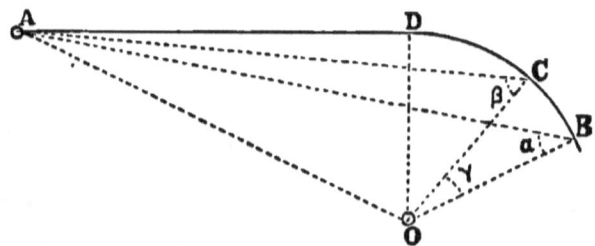

$$\text{Sin. } BAO = \frac{\sin.(\gamma + \alpha - \beta)}{\sqrt{1 - 2\frac{\sin.\beta}{\sin.\alpha}\cos.(\gamma + \alpha - \beta) + \frac{\sin.^2\beta}{\sin.^2\alpha}}}$$

$$\sin. DAO = \frac{\sin. BAO}{\sin.\alpha}$$

$$DOC = 90° + DAO - BAO - \gamma - \alpha.$$

When A is near C, and the curve has been run too far around, the following simple solution of this problem is given by Mr. J. S. Dunning, in the *Railroad Gazette* for March 17, 1882:

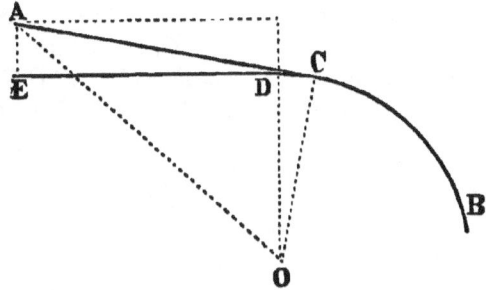

Measure DE and EA; then

$$\tan AOD = \frac{ED}{R + AE}$$

$$\cos COA = \frac{R \sin AOD}{ED}$$

$$DOC = COA - AOD.$$

In practice, however, AE is generally very small in comparison with ED, and it is then a sufficient approximation to turn the transit when at D on A, and measure the angle ADE, and make $DOC = ADE$.

Problem 10. To draw a tangent to two curves already located.
1st. When the curves are in opposite directions.

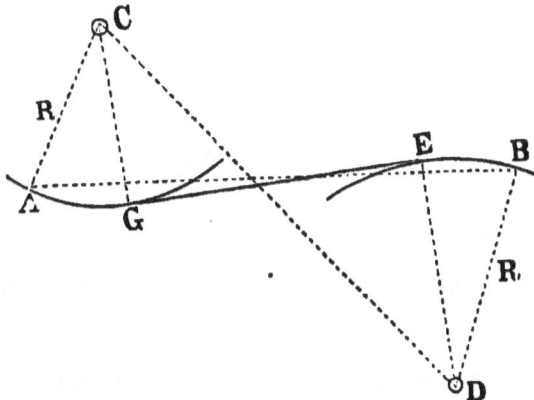

Stop both curves before getting to the tangent points. Observe BAC and ABD, and measure AB. In the triangle ABC, calculate CB, CBA and ACB (AC, AB and CAB being known). In the triangle BCD calculate CD, BDC and DCB (BC, BD and $CBD = CBA + ABD$, being known).

$$\text{Cos. } EDC = \frac{R + R'}{CD}.$$

$BDE = BDC - EDC$
$ACG = ACB - (EDC + DCB).$

2d. When the curves are in the same direction. This is an extension of Problem 4.

Problem 11. To substitute a compound curve for a simple one. ("Henck's Pocket Book," p. 59.)

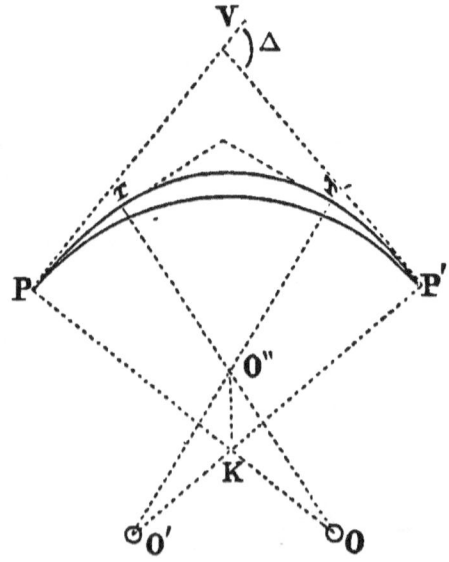

Let $PK = P'K = R$ and $PO = P'O = R'$ and $TO'' = T'O''$ $= R''$ and $POT = P'O'T = 2\theta$ and $TO''T' = 2\theta'$. Assume R' and θ'.

Then $\Delta = 4\theta + 2\theta'$ and $O'O'' : KO' :: \sin. O'KO'' : \sin KO''O'$,

or $R' - R'' : R' - R :: \sin. (\pi - \tfrac{1}{2}\Delta) : \sin. \theta'$.

$$\therefore R' - R'' = \frac{(R' - R)\sin. \tfrac{1}{2}\Delta}{\sin. \theta'}$$

Problem 12. To locate the second branch of a compound curve from a station on the first branch.

Let AB be the first branch of a compound curve and D its deflection angle, and let it be required to locate the second branch AB', whose deflection angle is D', from some station B on AB. Let n = the number of stations from A to B, and n' = number of stations from A to B'. Let $V = ABB'$; $(BAT = nD)$

$$V = \frac{n'(nD + n'D')}{n + n'}$$

(See "Henck's Pocket Book," p 61.)

Problem 13. To locate a tangent from an inaccessible point on a curve.

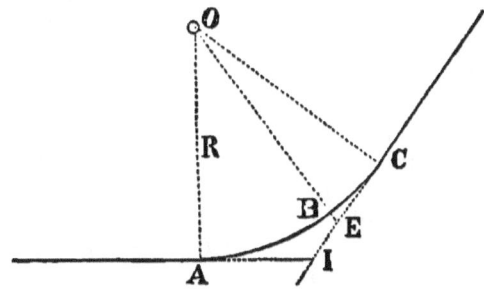

Let C be the inaccessible point. Run the line to a point B.

$$BE = R\left(\frac{1}{cos.\ COB} - 1\right).$$

$$BEC = 90° - COB.$$

Problem 14. To pass an obstacle on a curve.

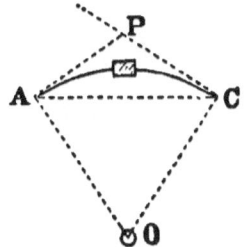

1st method. $AC = 2R \sin \tfrac{1}{2} AOC$.
2d   " $AP = R \tan \tfrac{1}{2} AOC$.

AOC must be assumed at such a value as, it is supposed, will carry the line beyond the obstacle.

Problem 15. To pass an obstacle on a tangent. ("Mifflin on Railway Curves," Prob 17.)

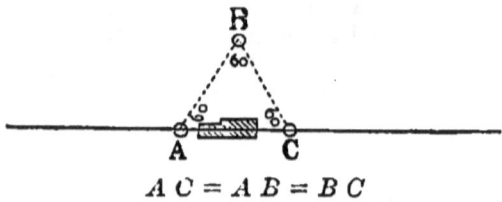

$AC = AB = BC$

Problem 16. To find the distance across a river in a preliminary survey.

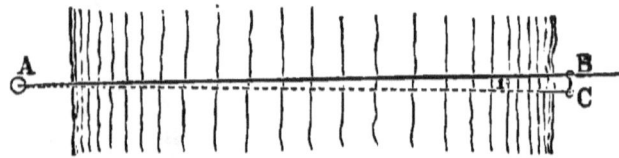

From A put in the plug B on the opposite side in the line which is being run. Then turn off one degree and put in the plug C. Measure the distance BC; then

$$AB = \frac{100\ BC}{1.75} \text{ or } \doteq 57.3\ BC.$$

Problem 17. To find the radius of a circular arc which shall successively touch three straight lines BD, DE and EC. (From "Rankine's Civil Engineering.")

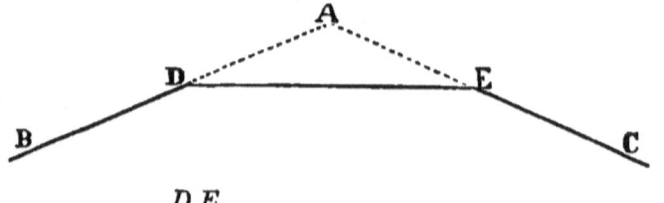

Radius $= \dfrac{DE}{\tan. \tfrac{1}{2} D + \tan. \tfrac{1}{2} E}$ ($D = ADE$ and $E = AED$.)

Problem 18. To connect two tangents with a curve of a given radius when the point of intersection is inaccessible. (From "Rankine's Civil Engineering.")

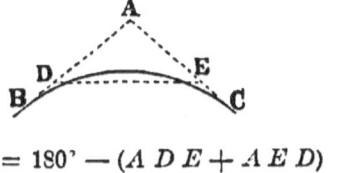

$$DAE = 180° - (ADE + AED)$$

$$AD = DE \frac{sin.\ AED}{sin.\ DAE}; \quad AE = DE \frac{sin.\ ADE}{sin.\ DAE}.$$

$$DB = R\ cotan.\ \frac{DAE}{2} - AD;\quad EC = R\ cotan.\ \frac{DAE}{2} - AE.$$

The transit points (marked *Tr. P.*) are called "plugs," and consist of stakes driven flush with the ground. They are guarded by a stake set on one side with the number turned toward the plug, and under it written (say) "3' off." All the other stakes should have their numbers turned toward the beginning of the line.

The following is the form of field-book: (From Shunk.)

Station	Distance	Deflection	Index	Tangent	Course	Mag. Course	Remarks
23 o	100				N.20°00'W.	N.20°05'W.	At sta. 24 + 50 commence a 4° curve to the L. for 35°12'.
24	50						
P C.+50 o	50	1°00'	1°00'				
25	100	2°00'	3°00'				
26	100	2°00'	5°00'				
27	100	2°00'	7°00'				
28 o	100	2°00'	16°00'	14°00'		N.34°07'W.	
29	100	2°00'	18°00'				

In locating in a mountainous country, it will generally be more conducive to economy to throw the heaviest grades in one part of the road, if possible. Auxilliary engines can then be used on those portions to assist in hauling the trains, so that the motive power can be more accurately adjusted to the size of the trains, and it will not be necessary for an engine to haul only a portion of its maximum load for a long distance. For example, the grades on the Pennsylvania Railroad in crossing the Allegheny Mountains, going west, are principally condensed into the portion between Altoona and Gallitzin, a distance of twelve miles, with a maximum of 100 feet per mile, while from Harrisburg to Altoona, a distance of 132 miles, the maximum is only 21 feet per mile.

In a line across the principal water-ways of a country, it will generally be better to make it undulating between the ridges and valleys than to go to great expense in making deep cuts and heavy fills to secure a level grade; for the interest on the increased capital account will amount to more than the extra cost of the motive power required to move the trains. The action of gravity, too, on the downward portions may save a part of the fuel which would be required on a level.

The maximum grade and minimum radius of curvature to be adopted depend on the judgment of the Chief Engineer, based on the probable amount of traffic the road will command. On the Hudson River Railroad the sharpest curve is 3 degrees; on the Pennsylvania it is 6 degrees, except in the mountain division, between Altoona and Gallitzin, where a $10\frac{1}{2}$ degree curve is employed in one instance. On the Callao, Lima & Oroya Railroad, across the Cordilleras in Peru, the sharpest is one of 120 metres radius, or a 14 degree curve. The maximum grade on the Pennsylvania Railroad is likewise on the mountain division, and amounts to $1\frac{9}{10}$ per cent. On the Oroya Railroad it is 4 per cent. On the St. Gothard Railroad, in Switzerland, the minimum radius is 280 metres, and the maximum grade $2\frac{1}{10}$ per cent.

The heaviest grade, however, is not usually combined with the sharpest curve. On the construction of the mountain division of the Pennsylvania Railroad in 1853, the maximum grade of $1\tfrac{8}{10}$ per cent. was only used on tangents, and on curves it was reduced .025 per cent. for each degree of curve. For example, the grade on the $10\tfrac{1}{2}$ degree curve is only 1.6375 per cent. On the Oroya Railroad, on curves of radii between 120 and 300 metres, 3 per cent. grade was allowed; but from curves of 300 metres radius to tangents 4 per cent. was the maximum.

It is probable that a reduction of one-tenth per hundred for each degree of curve would better suit the modern rolling stock. This may be put in form of an equation:

$$\text{Grade in ft. per 100 on a curve} = \text{max. grade on tang.} - \frac{\text{deg. of curve}}{10}$$

If this is preferred in terms of the radius, for

$$\frac{\text{degree of curve}}{10} \text{ write } \frac{573}{R}$$

Mr. H. G. McClellan finds that the resistance is not directly proportional to the degree of the curve, and deduced the following as the equation for curves of one metre gauge by experiment:

$$\text{Grade in feet per 100} = \text{max. grade on tangents} - \frac{213}{R-180}$$

the dimensions being expressed in feet.

The principal resistance on curves is due to the enforced parallelism of the axles in a truck, and is inversely proportional to a function of the radius, as expressed by these formulas. The use of the bogey truck, with a smaller wheel base than on European carriages, makes the American cars offer less resistance. For the former, therefore, a greater reduction would be necessary. It has been attempted to keep the axles always radial to the curve, but the devices do not seem to have been a practical success.

The other resistances on a curve are two : 1st, that due to the difference in length of the outer and inner rails, one wheel having to slip or slide by this amount when rigidly attached to the axle; and 2d, that due to the line of traction being on a chord of the curve, and not parallel to it. This latter resistance is greatest when the engine is starting from rest, and is counteracted by the centrifugal force. It can easily be calculated, and will be found to be just about neutralized when a car is 60 feet long and the train moves with a velocity of 15 miles per hour.

At the foot of heavy mountain grades, a reverse grade is sometimes put in for a short distance, to stop cars that may have become accidentally detached from the train, so that they shall not run into trains farther down the line.

When it is necessary to curve from one direction to another, a short piece of tangent should always be interposed to enable the proper elevation of the outer rail to be secured. On the St Gothard Railroad 40 metres is required in the specification. In the United States it is sometimes thought that two rails, or about 60 feet, is sufficient. Of course, the proper length depends on the sharpness of the curves, and the greatest permissible grade. (See remarks on a subsequent page in regard to curves of adjustment in connecting a curve and tangent; when they are used there is no necessity for a piece of tangent between reversed curves, although it is always better to have it.)

For a temporary track, with a gauge of 4 feet 8½ inches and ordinary engines, a curve of 187 feet radius may be made, round which the engine will easily go; on account of the danger of running off when going at great speed, and likewise on account of the great wear of both rails and tyres, such a curve should only be a temporary one. In *Engineering News* of Oct 2, 1880, a curve of 90 feet radius is mentioned, around which freight engines constantly go, this being likewise the minimum radius of the New York

Elevated Railroads, and in the "Transactions of the American Society of Civil Engineers," vol. VII., p. 107, a curve of 50 feet radius is mentioned, around which a very large traffic was conducted during the late war.

The final location having been made, the line is divided up into lengths of about one mile each, called sections, and a board placed on end at the dividing station, with the numbers of the sections on the sides. An estimate is made of the amount of earth-work and masonry on each section, and the road is advertised for contract. The contractors are each furnished with a printed copy of the quantities in each section, and allowed to take such notes as they require from the map and profile, and walk over the ground, the section boards guiding them to the different work.

CAMP EQUIPAGE.

Through the courtesy of Mr. E. T. D. Myers, the following list of articles required for "camping out" in preliminary and location surveys is inserted. It has been derived from an extensive experience:

One light wagon and horse, two two-horse wagons and horses, harness, etc., one saddle and bridle, five halters, five horse blankets, three wall tents and flies, 9 feet by 9 feet, with 3 feet walls and 8 feet high; one house tent, 8 feet by 8 feet, 8 feet high; four Indian rubber tent floors, twelve camp stools, two army combs and brushes, two lanterns, six candlesticks, one box of candles, one mess chest, one table, $4\frac{1}{2}$ by 3 feet; one table, 6 by 3 feet; twenty large tent pins, one small grindstone, one spade, two water buckets and dippers, two horse blankets, one cook's bucket, one box of soap, twelve boxes of blacking, two blacking brushes, four mattresses, one box of paints, six leather straps, two frying pans, one spider, one gridiron, one gill measure, one rolling pin, one weighing scale, one gal lon measure, one quart measure, one oven, one skillet, one pot, one tea kettle, one coffee pot, one pair of pot hooks, two tin pans, one bread pan, twelve tin cups, twelve tin plates, one tin washing pan, four tin basins, one pepper box, twelve coffee mugs, twelve plates, one dozen knives, forks, teaspoons and tablespoons, one large iron fork and one large iron spoon, one wooden tray, one molasses pitcher, one sugar dish, one butter dish, one two-gallon jug, one one-gallon jug, one coffee mill, one sifter, one dozen towels, one large dish, one coffee box, one sugar box, one lard box, one large pitcher, one large tub, one flat-iron (one medicine chest, with bitters, ginger, quinine, oil, oil of cloves, brandy, watch-maker's oil, laudanum, salve, bandages and lint, No. 6, calomel, horse fleam, lancet, hartshorn), smoking tobacco, pipes, chewing tobacco, one coil of small rope, two bunches of twine, one large twine needle, three small stoves, a half dozen papers of tacks, a half gross of matches, musquito bars, one dozen axes, three dozen transit books, four dozen level books, profile and mapping paper, tracing muslin, pens, knife, mouth glue, India ink, common ink, letter book, red ink, rubber for water colors, pencils, foolscap, letter paper, cartridge paper, chalk, red flannel, map cases of tin, box for stationery, drawing board, color cups and brushes, transit, level, rod and target, chain, tapes, transit rods, chain pins, triangles and rulers, drawing scales, hand axes and belts, brush hooks, small hatchet, two saws and saw files, six pounds of nails.

If provisions have to be carried, the following may be of use· According to the United States Army Regulations a ration is three-fourths of a pound of pork or bacon, or one and a fourth pounds of fresh or salt beef; eighteen ounces of bread or flour, or twelve ounces of hard bread, or one and one-fourth pounds of corn meal; and at the rate, to one hundred rations, of eight quarts of beans, or, in lieu thereof, twice per week, one hundred and fifty ounces of dessicated potatoes, and one hundred ounces of mixed vegetables; ten pounds of coffee, or, in lieu thereof, one and one-half pounds of tea; fifteen pounds of sugar, four quarts of vinegar, one pound of sperm candles, or one and one-fourth pounds of adamantine candles, or one and one-half pounds of tallow candles, four pounds of soap, and two quarts of salt.

The forage ration is fourteen pounds of hay and twelve pounds of oats, corn or barley; for mules, the same amount of hay, but only nine pounds of oats, corn or barley. For calculating the measure of the latter, one bushel of oats weighs 24, one bushel of corn weighs 56, and one bushel of barley weighs 48 pounds.

ADJUSTMENTS OF INSTRUMENTS.

The following is a convenient statement of the methods of adjusting the transit and level. It is taken from lectures delivered at the Rensselaer Polytechnic Institute, when the writer was a student there:

ADJUSTMENTS OF THE TRANSIT.

1st. To make the axis of the level tubes parallel to the graduated limb.

Test. Bring the bubbles to the centres of the tubes by the leveling screws, and turn the limb half-way around. The bubbles should remain at the centres.

Adjustment. Bring each bubble half-way back by the screws at the end of the tube.

In bringing the bubble to the centre of the tube by the leveling screws, it is well for the beginner to recollect that the opposite screws should be turned in pairs together, the thumbs approaching or receding from one another, and that the bubble will move in the direction of movement of the left thumb when it turns its screw.

2d. To make the line of collimation perpendicular to the axis of the trunnions.

Test a. Level the limb, and fix the cross-hairs on some well defined point on a level with the telescope and clamp.

b. Plunge the telescope and mark a point in the opposite direction now covered by the cross-hairs.

c. Unclamp and turn the limb until the cross-hairs cover the same point again, as at first, and clamp.

d. Plunge the telescope, and mark another point on a level with the telescope. This point should coincide with that marked in the process *b* above.

Adjustment. Move the vertical hair by the adjusting screws over one-fourth of the apparent distance from the last point marked, to the first. In making this adjustment, it is well to bear in mind that when the pins are inserted in the upper holes of the capstan-headed screws, pushing them from you makes the hair appear to move to the left.

3d. To make the axis of the trunnions parallel to the limb.

Test a. Level the limb carefully; fix the cross-hairs on some elevated point, and clamp.

b. Depress the telescope, and mark some low point covered by the cross-hairs.

c. Unclamp, and turn the limb half-way around; fix the cross-hairs on the elevated point and clamp.

d. Depress the telescope, and mark some low point covered by the cross-hairs ; this point should coincide with the former low point.

Adjustment. Lower the end of the axis opposite the second low point, by means of the adjusting screws on the standard.

4th. To centre the eye-piece. Although this is not necessary for accurate work, it is more agreeable to the eye.

Adjustment. Move the eye-piece by its adjusting screws until the intersection of the cross-hairs appears to be in the center of the field of view.

ADJUSTMENTS OF THE LEVEL.

1st. To make the line of collimation coincide with the axis of the telescope.

Test a. Fix the intersection of the cross-hairs on some well defined point by means of the leveling screws, and clamp.

b. Roll the telescope half over. The intersection should remain on the point.

Adjustment. Bring each cross hair half way back to the point by its adjusting screws.

2d. To bring the axis of the telescope and attached level into the same plane.

Test. Bring the bubble to the centre, and roll the telescope a little to one side and the other. The bubble should be stationary.

Adjustment. Bring the bubble back to the centre by the horizontal adjusting screws.

3d. To make the axis of the attached level parallel to the axis of the telescope.

Test. Bring the bubble to the centre and reverse the telescope end for end, in the Ys. The bubble should return to the centre.

Adjustment. Bring the bubble half-way back by the adjusting nuts at the end of the tube.

4th. To make the axis of the attached level perpendicular to the vertical axis.

Test. Bring the bubble to the centre over one pair of leveling screws, and reverse the bar. The bubble should remain at the centre.

Adjustment. Bring the bubble half-way back by the adjusting nuts at the end of the bar. In repeating the test, place the bar over the other pair of leveling screws.

5th. To centre the eye-piece. This is the same as for the transit.

6th. To make the object piece move parallel to the axis of the telescope.

Test a. Adjust the line of collimation on a *distant* object.

b. Fix the cross-hairs on a *very near* object, and roll the telescope half over. The intersection should remain on the point.

Adjustment. Bring each cross-hair half-way back by the adjusting screws of the object piece. In repeating the test, make use of the near object first.

CONSTRUCTION.

The road is divided up into lengths of about thirty miles, each of which is placed in charge of a "Principal Assistant Engineer." Each of these divisions is subdivided into lengths of about seven miles, and given in charge of an Assistant Engineer, whose party consists of a Rodman, Chainman and Axeman. The first work of the Assistant should be to retrace the line and test the bench marks. All plugs should be guarded, and a bench should be made at every culvert. There are various modes of guarding plugs—by intersecting lines, by distances from other plugs, or a combination of the two. The best method, where the ground admits of it, is by intersecting lines.

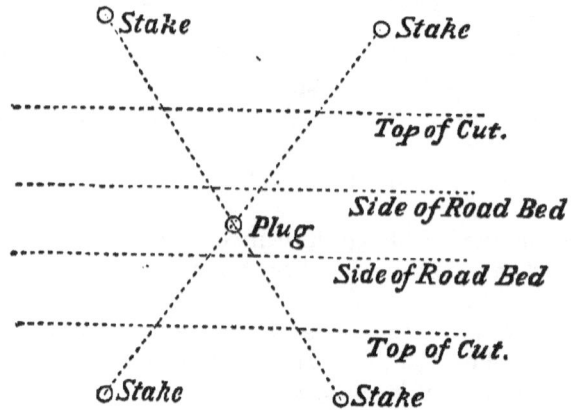

A part of a house, or the corner of a window or chimney, may often be substituted for one of the above stakes, for a foresight.

The note book may be kept in the form as on the following page.

Topography.	End of ridge, John Smith's house			Nail in root of tree.			Plug
Remarks............	Straight.						2° curve.
Total Angle........	11°20′			5°40′			
Reading...........	8°35′	7°40′	6°40′	2°50′	1°50′	0°50′	
Deflection...........	55′	1°00′	1°00′	1°00′	1°00′	0°50′	
Distance..........	86 + 92			84			81 + 17
Plug...............	P. T.			Tr. P.			P. C.
Station............	87	86	85	84	83	82	81

The advantage of this form of field-book is, that having first made and checked it in the office, there will be no more calculating of curves in the field, and so much less risk of error. It would always be necessary to run from the same end of the curve, and to use the same transit points; but this is no objection, as the plugs are all guarded, and it is as easy to set over one as another. Guarding all the plugs saves a great deal of trouble in re-running the line after grading, when it never measures the same as before, and it is difficult to run the old line without the *same* points to run from.

At the end of the transit note-book, a page should be devoted to each culvert, giving its station and a little plot of the stakes set, of course drawn roughly; and a little drawing of the culvert; also the level of the bridge seat and foundations, etc.

The staking out for excavation is done in several ways. In a tolerably flat or undulating country, it is generally done with the level; on steep hillsides, two rods are used, one ten feet long, and the other of any convenient length, divided into feet by different colors. That which is ten feet long is held horizontally by means of a hand-level laid upon it, with one end resting upon the ground, and the other against the shorter rod, which is held vertically. It is raised until it is horizontal, and the height read off the vertical rod by the Rodman, and noted by the Assistant on a piece of loose paper. He calculates where the slope runs out, and, having checked it on the ground, or made a closer approximation, enters it in a special field-book. The same principle governs the setting of the slope stakes with the level and level-rod. As a large part of the Assistant Engineer's work consists in setting slope-stakes, a more minute description is perhaps necessary. They are set opposite the centre line stakes, at the tops of the cuts and bottoms of the fills. If the slope is 1½ to 1 and the half width of the road-bed is called b, the horizontal distance of the slope-stake from the centre line is called x, and the height of the slope-stake above the sub-grade is called h; then

$$x = b + 1\tfrac{1}{2} h.$$

By assuming a value of x, measuring it out, and finding the corresponding value of h, with the level or the two rods, the values are substituted in the above equation: if both sides are the same, the assumed value of x was correct, and the stake should be driven in. If, however, the left-hand side is greater or less than the right-hand, the position of the stake should be moved toward or away from the centre line an estimated amount, and the process repeated of taking a new height with the level, and making a new calculation, until the two sides agree. After a little practice, it will not, usually, be necessary to make more than two trials. The following is the form of field-book used:

Station...	Distance..	Ground..	Grade..	Cut......	Fill...	L. D.....	L. C.....	R. C.....	R. D.....

The "L. D." signifies the distance out and the height of the ground where the cut or fill strikes the surface of the ground on the left-hand side, while "L. C." means the cutting or filling at the half width of the road-bed on the same side. (The object of finding this "L. C." or "R. C." is to facilitate calculating the areas of the cross-sections in the office.) Some engineers look upon the calculation of the point where the slope runs out, in the field, as a waste of time; and only take the transverse slope, being sure to take it far enough out. They then plot the cross-section, and take the distance to the slope-stake from the plot with a scale. They claim that this method is advantageous, too, because they always run out further than necessary for the slope, and if, afterward, as often happens, the slope will not stand, but slides out—a "slip"—they still have a record of the amount which slides by measuring to the top of the slide, while, too often, when such an accident occurs, the Assistant finds that he has no note of the slope of the ground beyond his stake, which has been carried away. When such an event occurs, it is better not to slope the cut further up, but to take away the earth at the level of the road-bed for some distance in, to catch any further slide before reaching the track, although the slope may be steeper than was intended.

In staking out with the level, it is well to have a number of sheets of paper, fastened together at the edges, for making trial calculations on; when one is covered with figures, it can be torn off and thrown away, exposing another. The cross-sections should be plotted in a permanent record book, to be kept in the office. The area of each should be calculated. For applying the prismoidal formula for calculating the cubic contents, it is requisite to know the middle cross-section between each two that are measured on the ground. The closest approximation to this is the following: Each cross-section is supposed to be transformed into another of equal area, but with a horizontal ground surface, and the depth at the centre of this new cross-section calculated. The depth of the middle section required is supposed to be equal to the mean of the two end "equivalent centre depths." From this depth the area of the middle section is obtained and substituted in the formula :

$$S = \frac{l}{6}(A + 4M + A'),$$

where A and A' are end areas and M is the middle area, and l is the distance of end stations apart. Tables have been constructed of "equivalent centre depths" for various areas, and other tables give the cubic contents at once, for a given length and given slopes, from the equivalent centre depths of the end sections.

Professor Rankine recommends a different method of finding the middle cross-section. Instead of finding the "equivalent centre depths" of the end-sections, and taking the mean for the centre depth of the middle-section, he assumes that it has a depth equal to the mean of the two actual centre-depths of the end-sections, and a slope equal to the harmonic mean of the slopes at the ends; from these data the middle section is calculated. In other words, if a and c are the slopes at the ends, the slope of the middle section is taken as $\frac{2ac}{a+c}$. If the slope of the ground changes in a cross-section, it is not clear how this method can be applied. When the cross-sections are measured at equal distances, as is generally the case in ground with easy slopes, it being sufficient to take them only at each even chain's length, the labor of calculation may be very much abridged by using the following formulas, also from Rankine:

When there is an even number of equidistant cross-sections:

Let $A, A', A'', \ldots A^m$ be the successive cross-sections and l their distance apart; then the total volume is

$$S = l \left\{ \frac{A}{2} + A' + A'' + \&c. \ldots \ldots + \frac{A^m}{2} \right\}.$$

When there is an odd number of equidistant cross-sections, $A, A', A'', \ldots A^n$

$$S = \frac{l}{3}\left\{A + 4A' + 2A'' + 4A''' + 2A'''' + \&c. \ldots + 2A^{n-2} + 4A^{n-1} + A^n\right\}.$$

When the cross-sections are bounded, at the top in cuts, or at the bottom in fills, only by lines joining the centre-stake to the side-stakes, a convenient form of the prismoidal formula for calculating, at once from the field notes of staking out, is as follows:

$$S = \frac{l}{6} \frac{(2H + H')(d_1 + d_1') + (2H' + H)(d_2 + d_2')}{2} - Tl$$

in which H and H' are the respective centre cuts or fills at the ends, plus the height of the triangle formed by the side slopes produced, with the road-bed; d_1, d_1'; d_2 and d_2' are the side distances, and T is the area of the above-mentioned triangle formed by the side-slopes, prolonged, with the road-bed. For the same width of road-bed and same side-slopes, the number to be added to the end centre-cuts or fills to obtain H and H' will be constant, as will be likewise T. The following is another form of the same equation, which may perhaps be preferred. It is given by Prof. Greene in *Engineering News*, Oct. 2, 1880 :

$$S = \frac{l}{6}\left\{(c_1 + \tfrac{1}{2}c_2)(d_1 + d_1') + (c_2 + \tfrac{1}{2}c_1)(d_2 + d_2') + \tfrac{1}{4}w(h_1 + h_1' + h_2 + h_2')\right\},$$

where c_1 and c_2 are the centre-cuts or fills, d_1, d_1', d_2 and d_2' are the side distances, w is the width of the road-bed, and h_1, h_1', h_2 and h_2' are the side heights.

Many engineers complain of the great labor of the calculations involved in using these formulas, although, in construction, there is usually ample time on rainy days to perform them, when there may be nothing else to do in the office. To save this labor, Mr. A. M. Wellington has lately published a work in which the quantities may be taken from graphical tables, to which the reader is referred.

At the top of cuts it is well to have a ditch made on the up-hill side to keep the slope from being washed down. Proper dimensions are:

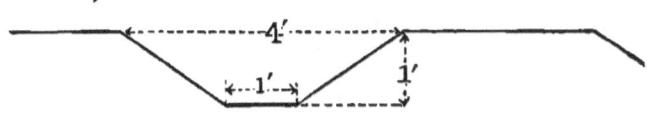

It should be placed about three feet from the edge of the slope.

It will sometimes be found that a cut passes through ground filled with springs, which come out on the sides of the slopes, washing them down, and continually filling up the ditches and increasing largely the amount of material to be removed. In such cases it is better to dig ditches in the face of the slope at right-angles to the centre-line, and fill them with broken stone, so as to form drains for the water, leading it to the side ditches. They may be made two feet wide and four feet deep, and about twenty feet apart, this distance however, of course varying with the amount of water delivered.

Estimates of the work done are taken up each month. It is important that all papers containing notes of the measurements should be preserved. Although these estimates are only intended as rough approximations, the measurements taken will often prove of service in following estimates.

CULVERTS.

For finding the proper water-way to give to culverts, the drainage area of the stream should be discovered if possible. Where county maps are obtainable, this can easily be measured from them. If the drainage area is small, it may often be estimated by walking round it. The water-way may then be calculated by the following formula of Mr. E. T. D. Myers:

$$A = c \sqrt{M,}$$

in which A is the area of the opening of the culvert in square feet, M is the drainage area in acres, and c is a variable co-efficient, depending on the country, and for which Mr. Myers recommends $1\frac{6}{10}$ in hilly, compact ground, and 1 in comparatively flat ground. In mountainous, rocky country, this value may often be raised to 4. Inquiry should be made of the neighboring people to learn the greatest height of floods in the stream, and the vertical dimensions of the water-way may be made equal to the flood height of the stream at the spot, although this is not necessary.

BOX CULVERTS.

Rule for laying out on the ground: Take the height of the top of the parapet from the height of the embankment at the centre, and with the remainder (considered as height of embankment) find the side distances with the level as in setting slope-stakes; then add 18 inches at each end, and if the height of the embankment exceeds 10 feet, add one inch on each end for every foot in height above the parapet.

The covering flags are one foot thick and the parapet one foot high, making two feet from top of abutments to top of parapet. For the thickness of abutments take $\frac{1}{4}$ the height of embankment on top of abutments, observing, however, that the abutments must never be less than two feet nor more than four feet thick. To determine the length of the wings, add the height of the opening to the thickness of the flags; one and a half times this sum, added to two feet, will give the distance from the end of opening to end of wing; the wing to be at right-angles to the drain, unless the latter be askew; then the wings to be parallel to the direction of the railroad. Instead of digging deep foundations, the method now employed is to put in a paving made of stones a foot deep, set up on edge, with a curb two feet deep at each face of the drain, and to start the walls on this paving. Should the fall of the drain not exceed 9 inches, make the pavement level, dropping the upper end 9 inches below the surface. Should the fall be greater, make a sufficient number of drops of 9 inches each in the length of the drain. At every drop place a cross-sill 2 feet deep; the wings and parapet to be of the same thickness as the abutments. The above rule was adopted on the construction of the Junction Railroad, of Philadelphia.

Box culverts are not usually made of a greater span than three feet. If more water-way is required, two openings are placed, each three feet wide, with a wall separating them, two feet thick.

OPEN CULVERTS.

These are generally made of two feet span, with walls two feet thick, with a depth of not more than three feet, founded on a paving, one foot thick.

CATTLE GUARDS.

These are often placed on each side of a public road crossing, when this takes place at grade. They are built like open culverts, with spans varying from three to five feet, and about three feet deep. Stringers 12" × 12" placed

3 feet 11 inches in the clear, support the rails, of a sufficient length to rest 5 feet on the solid wall and ground on each side. Two struts, $5'' \times 12''$, and 4 feet 6 inches long, and mortised into each stringer $3\frac{1}{2}$ inches, are placed about six inches further apart than the span of the opening, and the stringers are held to them by a rod one inch diameter, 6 feet 5 inches long, with square nuts and long flat washers, placed by each strut.

OPEN PASSAGE-WAYS.

These are made either with wing-walls or "T" abutments. When with wing-walls, the thickness at the base should be calculated like a retaining wall ($\frac{2}{7}$ the height). The wing-walls are usually placed at an angle of 45° with the centre line, that angle requiring least masonry. The coping then slopes down at the rate of 2.12 to 1, which is

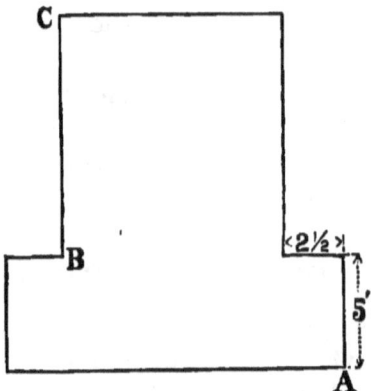

a good proportion for steps, if they are preferred. If the road is for a single track, the "T" abutment will be found more economical. The length of the "T" should be so calculated that the earth sloping down it at the rate of $1\frac{1}{2}$ to 1, and striking the back of the bridge-seat, and then one end, should just strike the ground at the corner of the bridge-seat, or as near it as the Engineer desires. For instance, suppose the distance from A to sub-grade is 12 feet; then

$$12 \times 1\tfrac{1}{3} = 5 + 2\tfrac{1}{4} + x, \text{ or } x = 10\tfrac{1}{4} = \text{length of } B\,C.$$

In staking out for a passage-way, always make the pit a foot larger all round than the foundation is intended to be, so that the quality of the masonry can be seen. The mason would prefer to fill up the entire pit. If the passage-way is on a curve, having decided where one face should come, turn off from the nearest plug the angle corresponding to the sub-chord to the face, and put in a plug; set up over this, and turn off the sub-chord to the face of the other abutment. Turn off right-angles from this last sub-chord at each of these plugs, and put in others outside of the pits for the mason to stretch his line by, for the faces of the abutments. Plugs should also be put on both sides of the last sub-chord produced, beyond the pits, to give the centre line of the bridge. This finishes the instrumental work, the other stakes being put in with a tape. A convenient way of doing this is as follows:

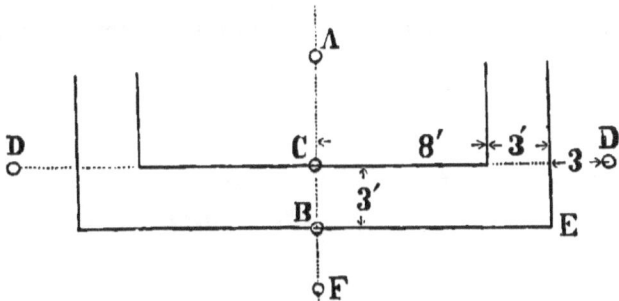

Let $A\,F$ be the centre line, marked with plugs at A and C, and let $C\,D$ be the face of the neat work. A stake is to be put in at the corner of the pit E, the pit being supposed to be 3 feet larger all round than the neat work.

Lining by eye, put in a stake B 3 feet from C. Then, with the ring of a tape at B and the 17-feet mark at D, take hold of the 14-feet mark and draw the tape tight; the 11-feet mark will give the point E.

It is well to give the mason a sketch on a piece of paper, giving all dimensions, drawn on the spot by eye without scale, and let him do his own marking out on the foundation. The pit is a sufficient guide for putting in the foundation. After setting the stakes for the pit, take levels at each one and note in the book; also note the depth of the pit before the masonry is begun, so that the cubic contents can be calculated. A level has also to be taken at the face before laying out the neat work, to give the height of the neat work to bridge-seat and for calculating the batter and span at the bottom. A 12-feet span bridge, 12 feet high, with a batter of one-half an inch to the foot, would be only 11 feet span at the base of the neat work.

STONE ARCHES.

Rankine's rule for the depth of the keystone in feet:

For a single arch, $\quad D = \sqrt{.12\,R}.$
For an arch in a series, $\quad D = \sqrt{.17\,R}.$

in which R is the radius at the crown in feet.

This is for circular or segmental arches. For elliptical arches, for R substitute $\dfrac{a^2}{b}$ when the earth is dry, or $\dfrac{4a^2}{b}$ when the earth is wet, a being the half-span, and b being the rise.

Trautwine's rule is:

$$D = \frac{\sqrt{R + \frac{1}{2}S}}{4} + .2 \text{ foot,}$$

in which R is the radius of the circle which will touch the crown and the springing lines, and S is the span.

Rankine gives as the thickness of the abutment from $\frac{1}{3}$ to $\frac{1}{5}$ of the radius at the crown (for abutment piers, $\frac{1}{2}$ the radius), and for the thickness of the piers $\frac{1}{8}$ to $\frac{1}{7}$ of the span. He says to make the masonry of the pier solid up to the point where a line from the centre of the arch to the extrados forms an angle of 45° with the vertical. Fill in the backing before striking the centres to such a height that $PQ = \sqrt{r'r - r^2}$, where r is the radius of the intrados, and r' is the radius of the extrados.

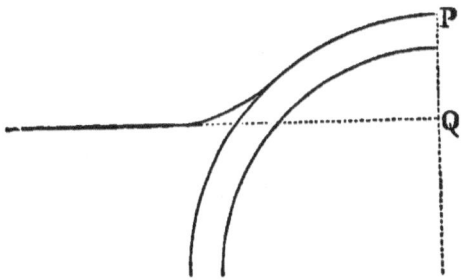

Trautwine gives for the thickness of the abutment at the springing line, when the height above the ground of this line is not more than $1\frac{1}{2}$ times the base,

$$= \frac{Rad.\ in\ ft.}{5} + \frac{rise\ in\ ft.}{10} + 2\ \text{feet}.$$

(See his "Pocket Book" for finding the thickness at the base.)

If the embankment over the arch is very high, or if the arch is the lining to a tunnel in earth, the proper form for the intrados is a geostatic arch. Rankine's approximate formula for this is:

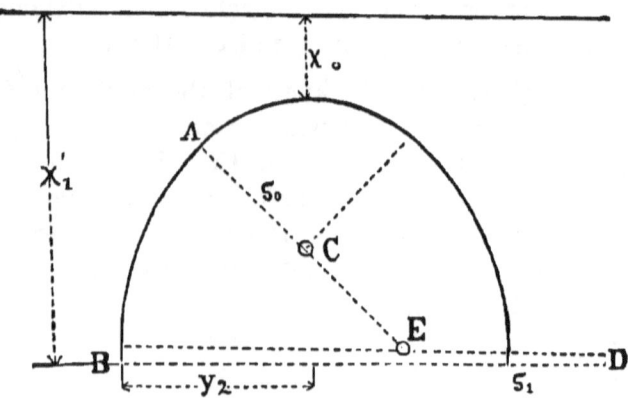

$$y_2 = \tfrac{1}{2}(x_1 - x_0) \sqrt[3]{\tfrac{x_1}{x_0}} \left\{ 1 - \tfrac{1}{30} \sqrt[3]{\tfrac{x_1}{x_0}} \right\}$$

In any given case x_1 and y_2 will be known, and we can calculate x_0 by trial; $x_1 - x_0$ will then be the rise of the arch, which we shall call a. The geostatic arch will approach a five-centre curve, which may be drawn as follows:

$$\text{Calculate } b = \tfrac{1}{2} y_2 + \tfrac{\tfrac{1}{4} y_2^2}{30\,a}.$$

$$\text{Then } \varsigma_0 = AC = \tfrac{a}{2}\left(1 + \tfrac{b^3}{a^3}\right) \text{ and } \varsigma_1 = BD = \tfrac{a}{2}\left(1 + \tfrac{a^3}{b^3}\right).$$

From these equations we obtain the centres C and D. About D with a radius $DE = \varsigma_1 - a$ describe a circular arc, and about C with a radius $CE = a - \varsigma_0$ describe another arc; the intersection of these at E will be the third centre.

When brick is plentiful, circular culverts are often employed. They require no foundation except for the face walls. For the thickness, one brick (nine inches) is sufficient for a span less than six feet. Add one more brick for each six feet more of span up to thirty feet.

The following are two cheap forms of centering:

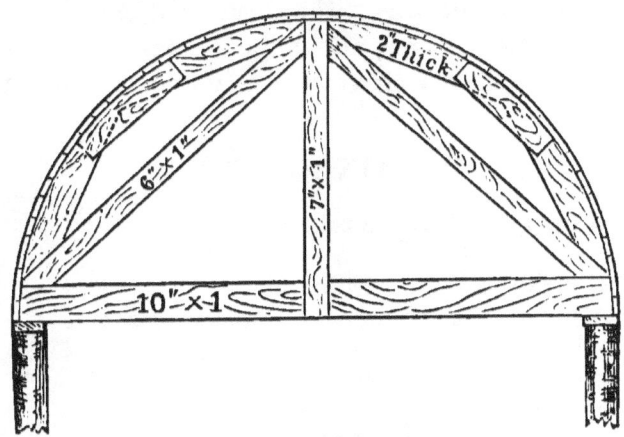

14-feet span. Frames 2½ feet apart.

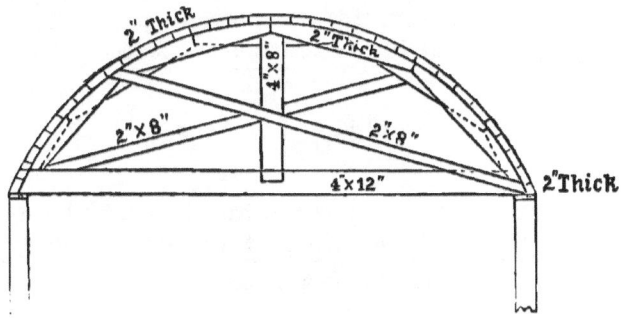

24-feet span, 8-feet rise. Frames 3½ feet apart.

Post mortised (by slight tenon) into chord. Arch pieces pinned together and halved in chord and post; braces spiked at ends; and at intersection with post a ½-inch bolt is used.

Centres should be removed from arches, unless laid in a very quick-setting mortar, within a few days after their completion. In stone arches the parapet should not be made too high, or it may be pushed over by the bank; it is well to proportion it like a retaining wall if more than one or two courses high. Some loose stones laid flatwise behind it will relieve the thrust of the earth.

In designing centres, allow $\frac{1}{300}$ of the span for settling of the arch, unless built very slowly and with great care.

RETAINING WALLS.

Let b = the breadth at the bottom = 1.
" h = the height.
" t = the thickness at the bottom.
" w = the weight of a unit of volume of masonry.
" w' = the weight of a unit of volume of earth.
" θ = the slope of the bank above the wall.
" φ = the angle of repose of earth.
" j = the inclination of the foundation pit to the horizon.
" q = a constant of safety

$$= \frac{\text{distance from middle of base to point where the line of resistance cuts base}}{t}$$

It is always between 0 and $\frac{1}{4}$.

Let q'

$$= \frac{\text{distance from the middle point of base to point where base is cut by a vertical line through the centre of gravity}}{\text{thickness at base.}}$$

It is always less than $\frac{1}{6}$.

$$\text{Let } n = \frac{\text{total weight of masonry}}{w\, h\, b\, t}$$

$$\text{Let } w_1 = w' \cos.\theta \; \frac{\cos.\theta - \sqrt{\cos.^2\theta - \cos.^2\varphi}}{\cos.\theta + \sqrt{\cos.^2\theta - \cos.^2\varphi}}$$

Rankine then gives:

$$\frac{t}{h} = \sqrt{\frac{w_1 \cos.\theta}{6\, n\, (q \pm q')\, w\, \cos.^2 j} + \left(\frac{w_1\, (q + \frac{1}{2})\, \sin.(\theta + j)}{4\, n\, (q \pm q')\, w\, \cos.^2 j}\right)^2} - \frac{w_1\, (q + \frac{1}{2})\, \sin.(\theta + j)}{4\, n\, (q \pm q')\, w\, \cos.^2 j}$$

If $\theta = 0$; $w_1 = w' \dfrac{1 - \sin\varphi}{1 + \sin\varphi}$.

If $\varphi = 35°$; $w_1 = .27\, w'$.
If $n = \frac{1}{3}$ and $j = 0$ in addition,

$$\frac{t}{h} = \sqrt{\frac{.27\, w'}{3(q \pm q')\, w}}.$$

If we suppose the wall to be just stable without the least excess of strength, $q = \frac{1}{2}$. It is customary, however, to give q a smaller value, so that there will be an excess of strength; otherwise the pressure would be concentrated at a point, and it would split off or crush. The English engineers make $q = .375$.

For q' we can assume the following values for walls of different heights:

Height of wall.	q'
20	.11
35	.14
55	.14
100	.16

For first-class masonry we may take $\dfrac{w}{w} = \dfrac{100}{165}$

For dry sand-stone rubble we may take " $= \dfrac{100}{120}$

Then if the wall is over 100 feet high, of first-class masonry, $\dfrac{t}{h} = .51$
" " " " dry rubble " .60
" " between 55 and 100 ft. high, first-class " .49
" " " " dry rubble " .58
If the wall is between 35 and 55 feet high, of first-class " .48
" " " " " dry rubble " .56
" " less than 20 " first-class " .45
" " " " dry rubble " .53

When the wall is rectangular in section:

If of first-class masonry, $\dfrac{t}{h} = .27$
" dry rubble " " .32

When $\theta = \varphi$, all the previous values of $\dfrac{t}{h}$ become more according to Rankine's formula.

The following are the rules used by different authorities:

In a discussion before the American Society of Civil Engineers ("Transactions," vol. 3, p. 75), a Canadian engineer was quoted as giving $\dfrac{t}{h} = \tfrac{2}{5}$ for first-class masonry laid in hydraulic cement. A rule used on the Pennsylvania Railroad is $\dfrac{t}{h} = \tfrac{3}{7}$. Rankine gives as the ordinary English rule, $\dfrac{t}{h} = .41$, and for a very safe rule $\dfrac{t}{h} = .48$.

Trautwine gives:

For rectangular walls of first-class masonry, $\dfrac{t}{h} = .35$
and of mortar rubble or brick   "   "   .40
and of dry rubble   "   "   .50

When the walls are offset at the back, he recommends a thickness at the base of about $\tfrac{1}{2}$ more, and at the top of $\tfrac{1}{2}$ less, containing the same amount of masonry. (See his book.)

It was the practice on the Pennsylvania Railroad to make the base $\tfrac{2}{7}$ of the height, and after carrying the back plumb for three or four feet to make a step, calculating the new thickness of wall at $\tfrac{2}{7}$ of the remaining height, and so continuing to step off to the top, where a thickness of three feet was given.

In railroads along a river bank, where the embankment slopes into the river, the slopes are "pitched" with stone about two feet long, laid upon the slope, at right-angles to it. They should start at the bottom in a trench dug about three feet below the surface of the ground.

TUNNELS.

A "heading" is first driven. This is about five or six feet high, and as wide as the nature of the material will allow. In earth it may only be three feet, while in solid rock it should be of the full width of the tunnel. In earth, it is generally driven at the bottom of the section of the tunnel; in this case, chambers are often excavated of the full size at intervals in the heading, and the work prosecuted from each in both directions until they meet. In solid rock, the heading should be at the top of the section of the tunnel, and the enlargement should be carried on as closely to the heading as possible, say within fifty feet, although where machine-drills are used, it may not be possible to keep so close.

When the two ends of the tunnel are so near the same level that the difference in their heights is not sufficient to secure a proper fall for drainage, a summit should be made in the tunnel. On the Mont Cenis tunnel a fall of .05 per 100 was considered sufficient, but Mr. B. H. Latrobe recommended .5 per 100. The St. Gothard has .1 per 100 (although level for 1,500 feet in the middle); Musconetcong, .15 per 100.

When shafts are sunk for the purpose of accelerating the work by having so many additional faces to work from, they should be filled up on the completion of the tunnel, as they interfere with the ventilation. It is found that the ventilation of the Hoosac tunnel is very bad since completion. It has a shaft in the middle which is left open, and acts well in removing the smoke from one end, but not from the other; the clear end depending, it is believed, on the direction of the wind.

When the tunnel is through earth or rotten rock, it will require to be arched. This is done with either brick or stone; in the London clay, which swells on being exposed to the air, it required a thickness of 54 inches of brickwork to withstand the pressure; 18 inches to two feet is the ordinary thickness. The Metropolitan Railroad, in London, goes under warehouses eighty feet high with a thickness of fourteen bricks, laid in cement, with a layer of concrete on top.

The proper form for the intrados of a tunnel through sand, or some such substance which acts only by its weight,

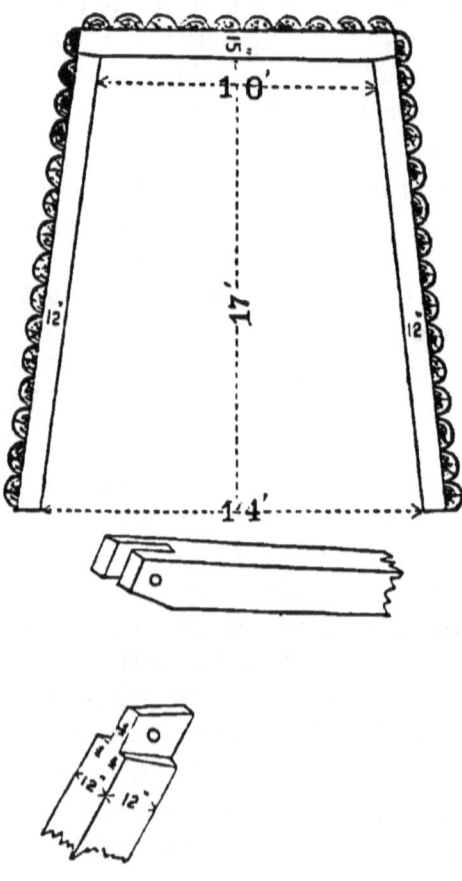

is the geostatic arch, to which a near approximation is made, when the load is infinite, in an ellipse with the semi-vertical axis double the semi-horizontal, or the rise equal to the span. In substances like the London clay, however, which swell on exposure to the air, a circular form is probably the best for the intrados. In soft material, the heading and the enlargement have to be timbered as they advance. The following was the method adopted on the Northwestern Virginia Railroad (see cut on preceding page):

Legs squared 12" and 17½' long. Cap, 12 feet long, 15 × 12 inches. Lagging half round, split out, about six inches thick, long enough to lie on two bents. After it was put in, the earth was rammed into the intervening space.

For the timbering on the Central Pacific Railroad, a longitudinal sill on each side, 12" × 12", carried the posts, 12" × 16", inclining outward at top, at intervals of 1½ to 5 feet. On these posts arches were made (polygons of seven sides) of three thicknesses of 5" × 12" plank, bolted with ¾ inch bolts. Width of sub-grade inside posts was 17 feet, and at springing line 19 feet. Height of crown above grade, 19 feet 9 inches. Split lagging on top, 2½ inches thick.

The timbering should be put in large enough for the masonry to be built inside; the earth is then tamped in above, the timbering sometimes remaining in, although it had better be taken out.

For running the line in rock tunnels, the transit points are made in the heading by driving wooden plugs in the roof, and centering them.

The excavation for the St. Gothard tunnel is 8 meters wide by 6 meters high, exclusive of space for the masonry. The heading was 2.4 meters high and 2.6 meters wide, kept about 200 or 250 meters ahead of the enlargement, at the top of the enlarged section. The enlargement is first made by cutting the place for the roof. About 200 or 250 meters further back, a cutting is made about 3 meters wide, down to the floor of the tunnel. About the same distance still further back, the whole section is excavated, and the remaining masonry put in. The heading of Clifton tunnel was 8 × 10 feet.

In the heading for a tunnel on the Great Western and Midland Railroads, at Bristol, thirty to forty shots were required to bring away the face, the holes being three feet six inches deep. They were exploded successively, beginning with the central holes, which were angled, and progressing to the outside ones. At the Mont Cenis, the machines were too long to allow of putting the first holes at an angle, and the first opening was made by putting down larger holes in the centre, which were not fired.

At the St. Gothard tunnel, the three central holes formed a triangle sixteen inches apart, converging to four inches at the bottom at a depth of $3\frac{1}{2}$ to 4 feet.

In the Musconetcong tunnel, a slope was made, instead of a shaft, 8 × 20 feet in the clear, at an angle of 30 degrees. Through earth it was timbered with collars 12 × 12 inches oak, 4 feet apart, centre to centre, supported by end and two middle props, lagged at the sides and above with chestnut " forepoling." Through rock the dimensions were 8 × 16 feet. Top headings were started in the tunnel 8 × 26 feet wide. Where the rock was disintegrated, collars of 15 inches oak, set 5 feet apart, were used, lagged above and sometimes at the sides, and supported either on legs or by hitches in the rock. These collars were sufficiently high to clear a two-foot ring of masonry, and about them packing was securely blocked in, up to the roof. The heading at the end of the tunnel was made 26 feet wide by 7 feet high. A heading through earth was made 8 feet at top and 10 feet at bottom, and 8 feet high, with oak collars and props of 12 to

15 inches round timber; sets placed 2½ to 2 feet apart, centre to centre, footed in very soft ground on six-inch sills, but ordinarily on three-inch foot-blocks. This information is obtained from Mr. Drinker's "Tunnelling," p. 221.

Tunnels are usually made 20 feet high from sub-grade to top, and 16 feet wide for single track and 26 feet for double track. Single track tunnels generally have vertical sides and a semi-circular top; and those for a double track have a cross-section composed of arcs of different circles tangent to each other, the upper half approximating a semi-circle whose centre is about nine feet above sub-grade. For various sections, and an immense amount of information on tunnelling, see Mr. H. S. Drinker's book. A common way of "bonding" the brick in tunnel linings is, to lay two consecutive courses of stretchers until the outer course falls behind the inner one just the thickness of a brick; the interval in the outer course is then filled in with stretchers, and a heading course follows. In a semi-circular arch of 16 feet span, these heading courses will occur at intervals of about 21 bricks. If the span is 26 feet, the heading courses would occur every 35th. As this is scarcely often enough, it is better to lay alternate courses of headers and stretchers, making the surfaces of the headers radial by thickening the mortar joint at the outer end. The successive nine-inch rings thus formed should be tied together with headers whenever their joints come in line, which will be about every 17th course.

It may be interesting to note that when the headings met in the St. Gothard Railroad tunnel, 14,920 feet long, the alignment was out 33 centimeters laterally, five centimeters vertically, and seven meters in length. (See *Engineering News* for June 5th, 1880).

BRIDGES.

Simple beam uniformly loaded, rectangular:
Let W = the breaking weight in pounds.
" b = the breadth of the beam in inches
" d = the depth of the beam in inches.
" L = the length of the beam in inches.
" S = a constant which has been determined by experiment.

The value of S is, for oak, 10,000; for white pine, 7,000; for wrought iron, 40,000; for cast iron, 30,000.

$$W = \frac{4}{3} S \frac{b\, d^2}{L}$$

Beam uniformly loaded, cylindrical:
Let r = the radius in inches, the other letters being as before

$$W = \frac{4\, S}{L}\, 3.1416\, r.$$

Beam uniformly loaded, I-shaped section:
If the flanges are of the same size, as they always are in rolled beams:

Let d = the depth of one of the flanges.
" d' = the depth of the connecting piece.
" A = the area of one of the flanges.
" A' = the area of the connecting piece.

$$W = \frac{4}{3}\frac{S}{L}\left\{4\,(d+d')\,A + \frac{d'^2\,(A+A')}{d'+2d}\right\}.$$

When the load is supported in the middle, instead of being uniformly distributed, the breaking load in each of the foregoing cases becomes only half as much.

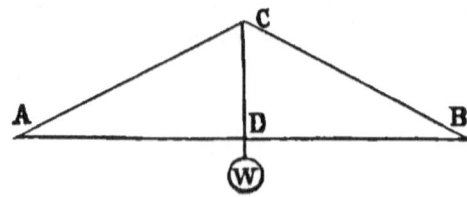

The king-post truss:

Strain on $CD = W$; strain on AC and $CB = \dfrac{W}{2}\dfrac{AC}{CD}$.

Strain on $AB = \dfrac{W}{2}\dfrac{AD}{CD}$.

If the load is uniformly distributed, it will produce the same effect as one-half the above load suspended in the middle. The beam is supposed to have no stiffness at D. Actually, AB is always made in one stick, and its stiffness will reduce the above values by an indeterminate amount

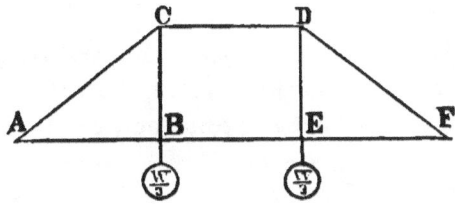

The queen-post truss:
$AB = BE = EF$.
Strain on CB and $DE = \frac{1}{3} W$.
Strain on AC and $DF = \frac{1}{3} W \times \dfrac{AC}{CB}$.

Strain on $CD = \frac{1}{3} W \times \dfrac{AB}{CB} = $ strain on AF.

These are the strains when loads of $\frac{1}{3} W$ are placed at B and E, or a total uniform load of W. In the latter case, the abutment at A has to sustain, in addition, the load on $\frac{1}{2} AB$, which, added to the resolved component of the strain on AC, produces a vertical strain of $\frac{1}{2} W$, as it ought.

If only the point B is loaded with $\frac{1}{3} W$, the portion which is transferred to the abutment F will produce a moment about D tending to break the joint across, if it is rigid. If, however, it is flexible, there will be a tendency for the joint to rise which is resisted by the rod DE (producing a strain upon it of $\frac{1}{3}W$) and the strain transferred to E, must be resisted by the transverse strength of the beam BF, calculated in the same way as the first case of the simple beam loaded in the middle (of a length BF, not AF). If braces are introduced in the directions BD and CE, the bridge becomes a Howe truss.

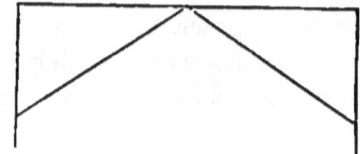

Strains are the same in this case as in the king-post truss, the tension on the straining beam, however, being converted into thrust against the abutments.

Strains are the same as in the queen-post truss.

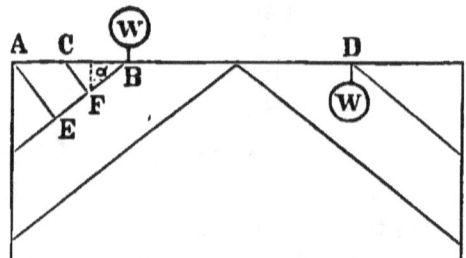

This is a combination of the king and queen post systems.

To prevent the point B from rising on the application of a weight at D, braces are often introduced at AE and CF. The force W, acting upward at B, produces a force equal to $W\sin\alpha$ in the direction of CF (CF being at right angles to FB) and acting at B, which has a tendency to break the beam transversely at F. The formula for the breaking weight of a beam fixed at one end and loaded at the other is $W = \frac{1}{6} S \frac{b\,d^2}{L}$. The pressure on CF is also $W \sin \alpha$, which produces a transverse strain in AB at C, which can be calculated in like manner.

The trussed girders of the following forms may be looked upon as king and queen post trusses inverted:

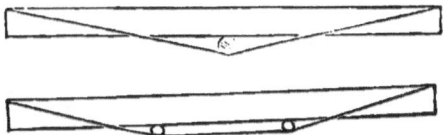

They are objectionable as being a combination of two systems, and it is impossible to tell just how much of the load will be borne by the beam acting as such, or by the rods transmitting a portion of the strain to compress the beam. It depends on the adjustment of the rods.

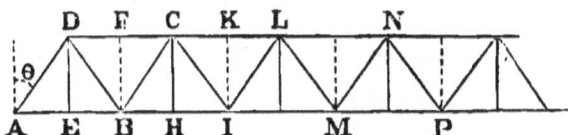

The "Warren girder," or "triangular truss:"

Let W = the weight on one panel = the weight on FC, due to uniform load or weight of bridge.
" W' = the weight on one panel = the weight on FC, due to the variable load.
" n = the number of panels, AE, EB, etc.

If the load is on the upper chord, struts FB, KI, etc., are introduced, and if on the lower chord, rods DE, CH, etc., so that each apex of the triangles will bear a load. The maximum strain on one of these struts or ties would be $W + W'$.

The maximum compression on $AD = \frac{n-1}{2}(W + W')\sec.\theta$.

The maximum compression on $DB = \frac{1}{n}W'\sec.\theta - \frac{n-3}{2}W\sec.\theta$, if a plus quantity; if it is a minus quantity, there will be no compression on DB.

The maximum tension on $DB = \frac{n-3}{2}W\sec.\theta + \frac{(n-2)(n-1)}{2n}W\sec.\theta$.

The maximum compression on $BC = \frac{n-5}{2}W\sec.\theta + \frac{(n-3)(n-2)}{2n}W'\sec.\theta$.

The maximum tension on $BC = \frac{3}{n}W'\sec.\theta - \frac{n-5}{2}W\sec.\theta$, if a plus quantity; if it is a minus quantity, there will be no tension on BC.

The maximum compression on $CI = \frac{6}{n}W'\sec.\theta - \frac{n-7}{2}W\sec.\theta$, if a plus quantity; if it is a minus quantity, there will be no compression on CI.

The maximum tension on $CI = \frac{n-7}{2}W\sec.\theta + \frac{(n-4)(n-3)}{2n}W'\sec.\theta$.

The maximum compression on $IL = \frac{n-9}{2}W\sec.\theta + \frac{(n-5)(n-4)}{2n}W'\sec.\theta$.

The maximum tension on $IL = \frac{10}{n}W'\sec.\theta - \frac{n-9}{2}W\sec.\theta$, if a plus quantity; if it is a minus quantity, there will be no tension on IL.; etc., etc., to the middle of the bridge, when the order will be reversed.

The strains on the chords are greatest when the bridge is uniformly loaded with its greatest load. We then have:

Tension on $AB = \frac{n-1}{2}(W + W')\tan.\theta$.

Tension on BI = tension on $AB + (n-4)(W+W')\tan.\theta$.
Tension on IM = tension on $BI + (n-8)(W+W')\tan.\theta$.
Tension on MP = tension on $IM + (n-12)(W+W')\tan.\theta$, etc., etc., etc.
Compression on $DC = (n-2)(W+W')\tan.\theta$.
Compression on CL = compression on $DC + (n-6)(W+W')\tan.\theta$.
Compression on LN = compression on $CL + (n-10)(W+W')\tan.\theta$, etc., etc., etc.

If the roadway is on the upper chord, the inclined piece, AD, in the last figure, is sometimes left out, and the bridge built as follows:

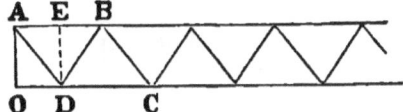

AD, DB, DE, etc., sustain the same amount of strain as before, but what was then compression is now tension, and *vice versa*. AB has the same amount of strain as AB in the other figure, and DC as DC the kinds being reversed as before. The part of the lower chord, DO, might, in this case, be dispensed with, were it not that it is of use in the lateral bracing, and must, therefore, be introduced.

The triangular truss is often built with two or more sets of different triangles forming the bracing; when of more than two, it is called a lattice truss. The strains on these can easily be represented in formulas as above, but are omitted for the sake of brevity. Each separate system of triangles has strains like the previous form, but independent of each other.

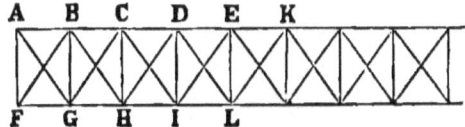

The Howe truss:

Let n = the number of panels.
" W = the weight on one panel due to the uniform load.
" W' = the weight on one panel due to the variable load.

Maximum compression on $FB = \dfrac{n-1}{2}(W+W')\dfrac{FB}{BG}$

Maximum compression on GC

$$= \left\{\dfrac{n-3}{2}W + \dfrac{(n-2)(n-1)}{2n}W'\right\}\dfrac{FB}{BG}$$

Maximum compresssion on HD

$$= \left\{\dfrac{n-5}{2}W + \dfrac{(n-3)(n-2)}{2n}W'\right\}\dfrac{FB}{BG}$$

Maximum compression on IE

$$= \left\{\dfrac{n-7}{2}W + \dfrac{(n-4)(n-3)}{2n}W'\right\}\dfrac{FB}{BG}$$

etc., etc. By continuing this process past the middle of the bridge, we obtain the maximum strains on the counter-braces, and when a strain becomes minus, it shows that beyond that point the counter-braces may be left out.

Maximum strain on BC

$$= \dfrac{n-1}{2}(W+W')\dfrac{FG}{BG}$$

= maximum strain on FG.

Maximum strain on CD

$$= \left\{\dfrac{(n-1)+(n-3)}{2}(W+W')\right\}\dfrac{FG}{BG}$$

= maximum strain on GH.

Maximum strain on DE

$$= \dfrac{(n-1)+(n-3)+(n-5)}{2}(W+W')\dfrac{FG}{BG}$$

= maximum strain on HI.

Maximum strain on EK

$$= \dfrac{(n-1)+(n-3)+(n-5)+(n-7)}{2}(W+W')\dfrac{FG}{BG}$$

= maximum strain on IL.

These calculations should be stopped at the middle panel; the other side will be the same.

The maximum strain on $B\ G = \dfrac{n-1}{2}(W + W')$ for a through bridge, or W' less for a deck bridge.

The maximum strain on $C\ H = \dfrac{n-3}{2}W + \dfrac{(n-2)(n-1)}{2n}W$ for a through bridge, or W' less for a deck bridge.

The maximum strain on $D\ I = \dfrac{n-5}{2}W + \dfrac{(n-3)(n-2)}{2n}W$ for a through bridge, or W' less for a deck bridge.

The Pratt truss has the same outline as the Howe truss, but the verticals are struts, and the diagonals ties. The strains would be the same as those for the Howe truss, except that $F\ G$, $G\ H$, etc., should be changed to $A\ B$, $B\ C$, etc., and $F'B$, $G\ C$, etc., to $A\ G$, $B\ H$, etc. In the Pratt truss, however, the strains on the verticals are as above for the through bridge, and W' more for the deck bridge.

Both trusses are sometimes made "double intersection," and the remarks on the Warren truss will apply also to these.

The Fink truss:

The through bridge, supposed to have sixteen panels.

Every part will receive its maximum strain when the bridge is uniformly loaded.

Let S = the span $A\ B$.
" D = the depth of truss $F\ E$.
" L = length of main suspending rod $C\ E$.
" L' = length of suspending rod $F\ G$.
" L'' = length of suspending rod $H\ I$.
" L''' = length of suspending rod $F\ N$.
" W = total weight of bridge and load.

The strain on center post $F\,E$ and end post $D\,B = \dfrac{W}{2}$

The strain on quarter post $H\,G = \dfrac{W}{4}$

The strain on eighth post $K\,I = \dfrac{W}{8}$

The strain on sixteenth post $M\,N$ = weight of $\frac{1}{4}$ chord $F\,K$ and $\frac{1}{4}$ of lateral braces.

The strain on suspending links $I\,L$ and $N\,O$ = weight of one panel with load.

The strain on suspending rod $E\,D = \dfrac{W}{4}\dfrac{L}{D}$

The strain on suspending rods $D\,G$ and $F\,G = \dfrac{W}{8}\dfrac{L'}{D}$

The strain on suspending rods $H\,I$ and $F\,I = \dfrac{W}{16}\dfrac{L'}{\frac{1}{2}D} = \dfrac{W}{8}\dfrac{L''}{D}$

The strain on suspending rods $K\,N$ and $F\,N = \dfrac{W}{32}\dfrac{L'''}{\frac{1}{4}D} = \dfrac{W}{16}\dfrac{L'''}{D}$

The strain on the chord is found by resolving the strains which are transmitted by the tension rods. Then

Strain on chord

$$= \dfrac{W}{4}\dfrac{\frac{1}{2}S}{D} + \dfrac{W}{8}\dfrac{\frac{1}{4}S}{D} + \dfrac{W}{16}\dfrac{\frac{1}{8}S}{\frac{1}{2}D} + \dfrac{W}{32}\dfrac{\frac{1}{16}S}{\frac{1}{4}D} = \dfrac{45}{256}\dfrac{WS}{D}$$

The deck bridge : In this case, the suspending links $N\,O$ and $I\,L$, etc., will be left out. The strain on the chord and tension rods and all the posts except the "sixteenth," $M\,N$, etc., will be the same as before. The sixteenth posts, $M\,N$, etc., will be strained by an amount equal to $\dfrac{W}{16}$

On the Pennsylvania Railroad, the greatest variable load that can come upon one track is supposed to be $1\frac{1}{2}$ tons per lineal foot, and all bridges are calculated for this. That is, the length of a panel multiplied by $1\frac{1}{2}$ will give the value of $2\,W$ in tons in the previous statement of the strains in the Howe truss bridge. The uniform load, or weight of one panel of bridge, 2 W, for bridges over 50 feet span, may be calculated from the empirical formula:

$$2W = l \left\{ \frac{1 + \frac{1}{5} S}{17 + \frac{1}{\sqrt{S}}} \right\}$$

in which l is the length of the panel in feet, S is the span in feet, and W is in tons.

After the design of the bridge is complete, the actual weight should be calculated from the drawings, and a correction made, if necessary. Theory seems to say that the weight per foot lineal, is proportional to the span, and if the span is large, the second term in the denominator may be neglected in comparison with the first, and the formula would accord with theory. However, the above seems to suit actual examples better; probably because in smaller spans the material, especially at the joints, must be disposed according to laws which differ from that of varying as the square of the span. For instance, in an iron bridge, if the chord is thickened at the joint to distribute the pressure of the pin over a larger area, the increase of weight due to the thickening is proportional to the number of panels, and is not necessarily a function of the span.

So when pins are proportioned to resist the crushing strain into the chord, they will be larger than is necessary for shearing, which gives an excess of material in the axis of the pin, which adds to the dead weight of the bridge, and is proportional to the number of panels, and likewise to the section of the chord, and independent to some extent of the span. The amount of material in the lateral and diagonal bracing will likewise increase with the distance apart of the trusses, even though the span remains the same.

The floor beams and track-stringers should be calculated for the greatest concentrated load which can come upon them, which is the weight on a pair of driving-wheels, and amounts to from 24,000 to 34,000 pounds. (Some of the later passenger engines even bring a load of 40,000 lbs. on a pair of drivers, and a uniform load of 3,700 lbs. per lineal foot for a length of 47 feet).

When the floor beams rest on a chord, its resistance must be calculated for the transverse strain so produced, in addition to the strain calculated on the supposition that the load is concentrated at the panel points. The English generally only allow the load to come on the truss at the panel points. In America, however, the advantage gained by supporting the engine on the floor beams, in case of a derailment, has been considered sufficient to retain the method of resting them on the chord. It may sometimes be well, if the trusses are far apart, to rest them on a saddle, so as

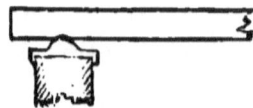

to insure their bearing equally on the chord, and not merely at the inside edge, when the floor beam deflects.

Care should be taken that the resolved strain from a brace to a chord should pass through the center of gravity of the section of the chord, and similarly from a strut to a tie. In a Howe bridge, by varying the size of the angle block, the axes of the brace and rod can easily be made to pass through this centre of gravity, that is, intersect there. This same point of intersection of the three strains in a Pratt bridge is the proper position for the pin. In reference to this pin, it may be remarked that in its manufacture the recommendations of Mr. Coleman Sellers in regard to axles should be observed ; that is to say, turn it approximately true with a fine feed, and finish with a coarse feed revolving rapidly. This obviates the effect of the wear of the tool, which would otherwise make the surface conical in stead of cylindrical. In Howe truss bridges, the upper chord is formed of several sticks of timber, which are made to act together by "keys" being notched into them.

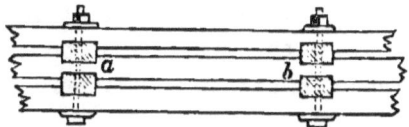

The distance apart of the keys, $a\ b$, should bear the same relation to the width of one of the sticks composing the chord as the length of one panel bears to the total width or depth (whichever is the least) of the chord. The depth of the notch is usually made ⅛ the width of one of the sticks composing the chord. If the keys are made of cast iron, they are made cross-shaped, the distance across the shorter arms being one-third the width of one of the sticks, plus the distance apart of the sticks, the latter being the amount necessary to let the rods pass between; the longer arms are made twice as long as the shorter arms, and the thickness of each is equal to the space between the sticks. A bolt passes through the chord and the keys.

The strength of the upper chord in the panel, and likewise of the braces, should be calculated by Gordon's formula.

For a long column, fixed at the ends, we have

$$P = \frac{f\,S}{1 + a\,\dfrac{l^2}{h^2}}$$

where P is the breaking weight in pounds, l is the length and h is the least external diameter, both expressed in inches, S is the cross-section in square inches, $a = \dfrac{1}{250}$ for pine and $f = 5{,}000$ for pine. If, as is usually done, the braces are bolted to the counter-braces where they cross them, their breaking strength must be calculated as for a column of one-half the length, with one end fixed, and the other rounded, which may be taken as a mean of the strengths of two columns of this half length, one with both ends fixed, and the other with both ends rounded. The formula, however, for both ends rounded is:

$$P = \frac{f\,S}{1 + 4\,a\dfrac{l^2}{h^2}}$$

It is only necessary to abut the several sticks of the upper chord against each other, without splicing them, making them break joints, however, so that there will not be more than one joint in a panel. The breaking weights calculated by the above formulas should be ten times the calculated maximum strains in a wooden Howe truss bridge, except for the rods, which should be able to bear six times the maximum calculated strain. This is called the "factor of safety." The reason is believed to be as follows: Mr. Fairbairn found by experiment that when the load upon a piece of iron was equal to one-third of the breaking weight, as found by gradually applying heavier loads, and this was successively applied and taken off, the piece did not seem to have its ultimate resistance injuriously affected. If, however, the strain exceeded about one-third of this breaking weight, the piece would break after a sufficiently great number of applications of the load. It was also found by experiment, by an English Parliamentary Commission to test the value of iron for railroad construction, that a load suddenly applied to a bridge, as when an engine came rapidly upon it, produced double the amount of strain that would be produced by it if it were gradually applied. A multiplication of these two factors gives one-sixth of the breaking weight as found by a load gradually increased to breaking, as the greatest allowable strain to come upon a bridge which is subject to a rapid application of the load, applied many times. For wood, this constant is made one-tenth, since its strength varies very much in different specimens, and the experiments from which its strength has been found were made with selected pieces, of small size.

Instead of taking one-sixth of the ultimate strength for the safe load, it would no doubt be better to take one-half of the so-called "limit of elasticity," or the point at which successive applications of the same load produce an increasing set. Experiments on the strength of materials however, have been heretofore directed rather to the ultimate strength, and this is what is recorded in the published tables. It may be observed, too, that the experiments from which the fact has been derived, of more than one-third of the breaking weight finally producing rupture, refer only to pieces strained in tension. It is probable that in compression a piece could sustain a repeated application of a greater proportion of the ultimate static load without rupture. Experiments are needed, however, to certainly establish this. In the Howe and Pratt bridges, each piece is only strained in one direction, either in tension or compression. In the Warren truss, however, the same piece may be strained sometimes in one way and sometimes in the other. The experiments of Wöhler tend to confirm the correctness of an American practice, that in this case such a piece should be proportioned to resist the maximum amount of compression; and should have, *in addition*, sufficient material to resist the maximum tension. The apparent advantage which the Warren has of requiring less pieces than the Pratt, and making one piece do double duty, is thus seen to be only apparent and not real, since the braces must contain an additional amount of material. This will perhaps explain why the Warren truss has never been so popular in the United States as it has been abroad.

In a Howe truss bridge the lower chord sticks have to be spliced, and the very common practice is to make the splice too weak.

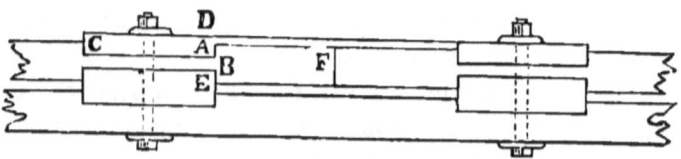

This is, perhaps, the commonest form of such splice.

The notch AB is generally too small, and bridges are often seen in which it is crushed. The proper relation between the parts of this splice is the following:

Let c = the crushing resistance of the wood = 5,000 lbs. per square inch for pine.
" t = the tensile resistance of the wood = 10,000 lbs. per square inch for pine.
" s = the shearing resistance of the wood = 600 lbs. per square inch for oak and 400 lbs. for pine.

Then

$$c \times 2AB = t \times BE = s \times 2AC = s \times 2BF.$$

For pine, we see that AB will equal BE and just two-thirds of the chord will have to be cut away. The splice-piece and keys are generally made of oak, or other hard wood. This is objectionable, since wood rots much more rapidly when it is in contact with a different species. A much better splice for the lower chord is that recommended by the late Mr. B. H. Latrobe, in the *Railroad Gazette*, vol. 10, p. 501. A flat iron link is set into the chord on each side. As a very small breadth is necessary to give the iron sufficient shearing resistance corresponding to the part AC, in the above wooden splice, sufficient space is left on the chord to put two or more additional surfaces on the link to withstand the crushing force

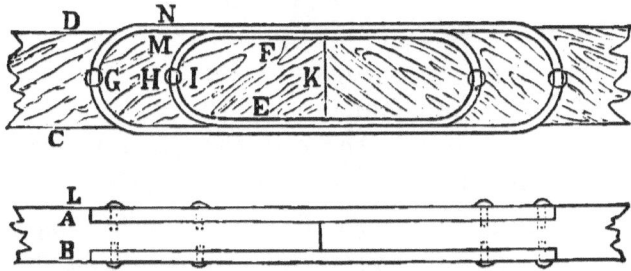

The following are the proper relations for determining the sizes:

$A B \times C D \times$ tensile resistance of the wood, should equal

$(C D + E F) \times A L \times 2 \times$ crushing resistance of the wood, and should also equal

$(G H \times C D + I K \times E F) \times 2 \times$ shearing resistance of the wood, and should also equal

$A L \times M N \times 8 \times$ tensile resistance of the iron bands.

Other forms of splices may be seen in the *Railroad Gazette*, vol. 10, p. 573, in an article by Mr. Geo. L. Vose.

The amount that the angle blocks are notched into the chord is a point to which not sufficient attention has been given in many existing bridges. They should have sufficient bearing surface to resist the crushing strain transmitted from the brace to the chord. It fortunately happens that this is greatest at the ends, where there is an excess of material in the chords, as they are made of the same section throughout, generally, in wooden bridges. The necessary amount of notching, of course, diminishes to the middle.

The nuts at the ends of the rods in a Howe truss bridge rest on wrought-iron pieces, which act the part of washers to distribute the load over a considerable surface of the wood, to prevent crushing of the fibre. Cast-iron "tubes" are generally used to transmit the strain from the rod to the angle block. They are ordered about a half-inch shorter than the depth of the chord, to allow for shrinkage in the latter. The rods are generally ordered of such a length that they will project beyond the nut a distance equal to half the diameter of the rod. The joints of the chords, especially those of the lower chord, should be painted with red lead.

Provision should be made in the angle blocks, when of cast iron, for the inserting of dowel pins to hold the braces and counters in place until the rods are screwed up. They may be made of short lengths of iron rods three-quarters of an inch in diameter and six inches long. If wooden angle blocks are used, which, however, is objectionable, nailing the braces to them is sufficient.

In Howe truss bridges the parts $A\,F$, $A\,B$ and $A\,G$ in the figure on page 57 being without direct strain, can be left out. This is seldom done, however, since the counter-brace serves to stiffen the main brace by shortening its length as a column. If the bridge is a deck bridge, these pieces serve to carry the track over the space $A\,B$ to the abutment. $A\,F$ is then made of wooden struts strong enough to resist the half load on a panel, and the chords are held against the struts by rods.

Lateral and diagonal bracing is recommended to be of sufficient strength to resist a pressure of wind of 30 pounds per square foot of truss and train, using a factor of safety of one-sixth. In the tropics, where hurricanes occur, it should be calculated for a pressure of at least 50 pounds per square foot. Such has been the prevailing practice in the United States. Mr. C. Shaler Smith has, however, noted a pressure of 84 pounds at the St. Charles bridge, and a velocity of 138 miles per hour has been likewise noted. The latter reduced to pressure by Smeaton's rule $p = \dfrac{v^2}{203\frac{1}{4}}$ gives a pressure of 93 pounds.

It is therefore recommended that in exposed situations it should be calculated for 100 pounds per square foot of bridge and train, and regarding, in an uncovered bridge, that one side does not shield the other. Considering however, the rarity of such hurricanes, it may perhaps, be allowed to use a factor of safety of 4. It should be stated however, that some modern experiments seem to indicate that Smeaton's rule gives pressures about double what they should be. See Mr. Wellington's experiments, Trans. Am. Soc. of Civil Engineers, vol. 8, p. 45.

The lateral bracing is generally made of the Howe or Pratt truss form, and since the strain may come upon it in either direction, the braces and counters are of the same size, acting alternately as one or the other, according to the direction of the wind.

Attention should be paid to the method of transferring the strains from the angle blocks of the lateral bracing to the chords, and from the diagonal bracing to the rods. It may often be necessary to use thicker sticks in the lower chord to resist this strain, if it is requisite to notch the angle blocks in, to obtain sufficient bearing surface. For small spans, however, the bearing of the angle blocks on the rods, which bear on the chords, may be sufficient to resist this strain without notching.

In erecting Howe truss bridges, after the pieces of a truss are asembled on the false works in their proper positions, the rods are finally screwed up from both ends toward the middle. In Pratt truss bridges, the main braces are first screwed tight, those that are symmetrically placed on each side of the centre being screwed up together, and advancing from the ends toward the middle. The counters are afterwards screwed up, their adjustment being determined by the note which is given out when they are struck with a hammer.

On the Pennsylvania Railroad the floor system designed

by Mr. Joseph M. Wilson consists of white oak floor beams 7 inches wide by 12 inches deep, placed 15 inches apart, centre to centre, resting on the chords and notched half an inch on them. The rails are spiked directly to these floor beams, and, outside of them, longitudinal guard timbers of white oak 6 inches square are placed, notched one inch on the floor beams, bolted to them at every fourth beam with one inch bolts with flat heads, and spiked at the intermediate ones. The chord must be calculated to resist the transverse strain thrown upon it by these floor beams, in addition to the strain obtained from the strain sheet. On the inside of each rail, a guard rail is likewise fastened to the floor beams, close to the traffic rail.

For bridges of twenty-feet span, wrought-iron built I-beams of 12 inches depth are used, two to each rail. On these cross-ties, 6 × 8, 7½ feet long, are notched one inch, spaced two feet apart; a stringer, 6 × 12, notched half an inch, and four feet longer than the iron beams, rests upon the cross-ties, and supports the rails.

Back-wall plates, 6 × 12: The coping should be strong enough to resist the pressure without wall-plates, using a cast-iron shoe 1½ inches thick, which should be large enough to distribute the load so that not more than 25,000 pounds per square foot come on the coping.

Bridges are always built with the upper surface slightly convex, so that when they are loaded the track shall not be below a horizontal line joining the extremities of the span. This rise is called the camber, and it is put into a bridge by making the upper chord longer than the lower one, and likewise making the main bracing (if not adjustable) to suit this additional length.

The additional length is $= \frac{(a + a') S}{E}$ in which a and a' are the maximum safe compressive and tensile strains on the material, per square inch, S is the span, and E is the modulus of elasticity. For wood $a + a'$ may be taken at 2,000 and E at 1,600,000; for iron $a + a'$ is about 19,000 and E is 22,000,000. The additional length of chord thus found must be divided up among all the panels. By making the substitutions it is seen that the upper chord of a wooden bridge must be in the proportion of $1\frac{1}{2}$ inches longer than the lower one for each 100 feet of span; and of an iron bridge one inch longer. If the height of truss is one-eighth of the span, the additional length of the chord will equal the camber. There is some reason for thinking that the most economical height for a truss (regarding only its resistance to the vertical load, and not to the wind pressure), is from one-eighth to one-tenth the span; the most economical angle for the inclined braces is 45° with the vertical. (See appendix for a discussion of the most economical height). To obtain both of these advantages with a small panel length, it is necessary in long spans to have two or more systems of braces and the bridge is then called double or treble, etc., intersection. It would perhaps, however, be productive of the greatest economy to use only one system of bracing with the above mentioned height and angle, and use smaller trusses to span the distance between panel points. The material would then be concentrated in a few large parts. It should be observed however, that many engineers, perhaps most of them, would disagree with the author on this point, although he thinks the prevailing tendency of modern bridge engineering is in that direction.

RIVETTING.

In bridge construction rivets have, according to their positions, two functions to perform; first, to transmit a tensile strain from one piece to another, and second to cause two or more pieces in compression to act together, so as to give a united resistance. The first function is the only one that is required in boiler construction, and the rules that are generally given for rivetting are derived from experiments made by boiler makers. When two pieces in tension are fastened with rivets in bridge construction, they are always "double rivetted," or have a cover plate rivetted on each side of the plates that are connected. The following is a sketch of this mode of construction.

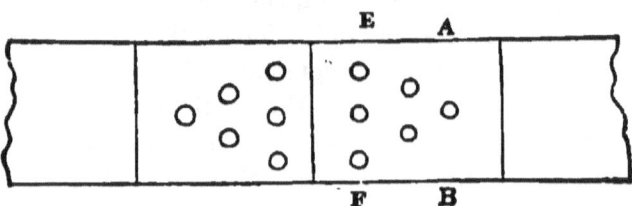

This joint may give way in either of the following manners: 1st. By tearing across the plate at AB. 2d. By tearing across the two cover plates at EF. 3d. By shearing the rivets. 4th. By the rivets crushing into the plate.

Let $t =$ the thickness of the plate in inches.
" $t' =$ " " " each cover plate in inches.
" $w =$ " width of the plates.
" $d =$ " diameter of the rivets.
" $n =$ " number of rivets in the row nearest the joint, as row EF.
" $T =$ " tenacity of the iron = 60,000 lbs. per square inch. (It should elongate fifteen per cent. before breaking).
" $S =$ " shearing resistance of the iron = 40,000 lbs. per square inch.
" $C =$ " crushing resistance of the iron = 20,000 lbs. per square inch.

This last value is derived from experiments on pin connections. Experiments on rivetted connections seem to make it as high as 80,000. It is probable however, that the friction between the rivet head and the plate increased the strength in the experiments, so that the rivet did not bear against the plate with its length only. In bridge construction the vibration of the bridge would probably render unreliable this friction, so that it is better to take a smaller value and use more rivets.

The equations for equal strength are

$$T(w-d)t = 2 T(w-nd) t' = \frac{\pi d^2}{4} 2 S \frac{n}{2} (n+1) = t d \frac{n}{2}$$

$$(n+1) C = 2 d \frac{n}{2} (n+1) t' C.$$

For ordinary sized plates, where more than one rivet is used, rivets of a sufficient size to resist shearing, will contain more than enough bearing surface. For pin connections it is otherwise.

We then have for rivets the equations:

$$T(w-d)t = 2 T(w-nd) t' = \frac{\pi d^2}{2} 2 S \frac{n}{2} (n+1),$$

from which we deduce

$$n^2 + n = 2 t \frac{w-d}{d^2}$$

and

$$t' = \tfrac{1}{4} \frac{w-d}{w-nd} \cdot t.$$

w, d, and t are the values usually given. Ordinary values of d range from $\tfrac{3}{4}$ to 1 inch, and such a value between these limits is chosen as will make $d = 1\tfrac{1}{2} t$ to $2 t$ if possible.

For pin connections, where bearing surface is of more importance than shearing surface, the proper equation is, n being equal to one,

$$T(w-d)t = d t \frac{n}{2} (n+1) C$$

from which we deduce

$$d = \tfrac{1}{4} w.$$

The results of experiments in England, according to Mr. Berkeley, show that the following are the relations of the dimensions for the strongest form.

Bar width $= 1$; diameter of pin $= \frac{3}{4}$; sum of sides of eye $= 1\frac{1}{4}$; end of eye $= 1$; radius of shoulder $= 1\frac{1}{2}$ (to 2).

According to experiments by Mr. C. Shaler Smith (Trans. Amer. Soc. of Civil Engineers, Sept. 1877) it seems that when several bars of varying widths come upon the same pin, the latter should be of a diameter not less than two-thirds of the width of the largest bar, the section across the side of the eye should be 1.33 for hammered, and 1.5 the width of the bar for "weldless" eyes; and the maximum thickness of these bars should not be more than one-fourth of their width. The metal section across the eyes in the smaller bars, and *their* thickness, should bear the following relation to the width of the bar.

Diameter of pin divided by the width of the bar.	HAMMERED EYES.		WELDLESS EYES.	
	Section across side of eye.	Max. thickness of bar.	Section across side of eye.	Max. thickness of bar.
$\frac{3}{4}$	$1\frac{1}{3}$	.21	1.5	.21
$\frac{7}{8}$	$1\frac{1}{3}$	.25	1.5	.25
1	$1\frac{1}{2}$	.38	1.5	.38
$1\frac{1}{8}$	$1\frac{1}{2}$	.54	1.6	.54
$1\frac{1}{4}$	....	....	1.7	.59
$1\frac{1}{2}$	$1\frac{3}{4}$	.70	1.85	.70
$1\frac{3}{4}$	$1\frac{3}{4}$	.88	2.	.88
2	$1\frac{3}{4}$	1.08	2.25	1.08

If the bar is "upset" at the ends, for t in the second member of the equation substitute $t' =$ the thickness of the upset part. In iron bridges the American practice is to make the portions which bear tension of eye bars. The parts in compression are universally made, when wrought iron is used, of members "built up" of several pieces of various sections, rivetted together. When compression members are made up in this way, the centre of gravity of the united section should coincide with the centre of figure; this can generally be accomplished by a manipulation of the parts in the design.

The following are some sections of upper chords.

For calculating the strength, Rankine's formula, in the following form, for columns fixed at the ends, should be employed.

$$\frac{P}{S} = \frac{36,000}{1 + a\frac{l^2}{r^2}}$$

where P is the breaking load in pounds, S is the cross-section in square inches, l is the length in inches, r is the "least radius of gyration" in inches, and a is constant for the same shape. (If the ends are rounded, or consist of eyes resting on pins, $4\,a$ should be substituted for a.) The column is then treated as a whole, and in order that it may so act, the rivets connecting the parts should be so spaced that their distance apart bears the same proportion to the thickness of either of the parts connected as the least diameter of the chord (or column) bears to its length between panels. A factor of safety of one-sixth should be employed. The value of a in the equation given by Prof. Rankine is usually taken at $\frac{1}{36000}$, a value derived from experiments on rectangular columns. A sufficient number of experiments on wrought iron in various shapes has not yet been made, and such experiments are perhaps the most important want that exists in the engineering profession, since "built" columns have come into universal use. Meanwhile the following values of a are offered for use until such time as an exhaustive series of experiments gives more information on the subject. The values of r^2 can be readily calculated from the formulas in the table.

Mr. Burr's formula for the strength of columns is

$$\frac{P}{S} = \frac{a}{\left(\frac{l}{r}\right)^b}$$

For "Phœnix" columns $a = 65,354$, and $b = .138$; for hollow cylindrical wrought iron columns $a = 116,390$ and $b = .317$.

The value of r in the formula, for any irregular section, if it is symmetrical with reference to two axes at right angles to each other, may be found by experiment thus. Draw the section on a large scale on cardboard; cut it out; suspend it by a pin thrust through it, near the edge, on a

Figure	a
	$\frac{1}{30104}$
	$27\frac{1}{2}77$
	$\frac{1}{5707}$
	$\frac{1}{64003}$
$+ 2ik\{\frac{1}{6}k^2 + 2(\frac{1}{2}d_3 + d_1 + g + \frac{1}{2}k)^2\}$	$\frac{1}{45715}$
$\frac{1}{2}b)^2$	$\frac{1}{44119}$
	$\frac{1}{13470}$
	$\frac{1}{25005}$

$$\tfrac{1}{12}(b-l)^2 + \tfrac{1}{6}\tfrac{bl(b+l)}{b+l-t} - \tfrac{1}{4}\tfrac{b^2l^2}{(b+l-t)^2}$$

$$\frac{\tfrac{\pi}{4}(R^2-r^2)(R^2+r^2)+2lb\left\{\tfrac{1}{12}(l^2+b^2)+\left(R+\tfrac{l}{2}\right)^2\right\}}{\pi(R^2-r^2)+4lb}$$

About AB $\left\{\begin{array}{l}\tfrac{1}{12}(d_2b_2{}^2+2d_1b^3+2ym^3)+i(g+k)(\tfrac{1}{4}i^2+2im+m^2)\\ \quad d_2b_2+2d_1b+2g(m+2i)+4ki\end{array}\right.$

" CD $\left\{bd_1{}^2(\tfrac{1}{3}d_1+d_3)+\tfrac{1}{4}d_3{}^2(\tfrac{1}{3}b_2d_2+bd_1)+(m+2i)g\{\tfrac{1}{3}g^2+2(\tfrac{1}{3}d_3+d_1+\tfrac{1}{3}g)^2\}+2ik\{\tfrac{1}{3}k^2+2(\tfrac{1}{3}d_3+d_1+\right.$
$\qquad b_2d_2+2bd_1+2g(m+2i)+4ki$

About AB $\left\{\begin{array}{l}\tfrac{lb^3}{12}+\tfrac{2l_1b_1{}^2}{3}-(l-2t)(b-t_1)\{\tfrac{1}{3}(b^2+bt_1+t_1{}^2)+n(a+b+t_1)\}+lb(a+\tfrac{1}{2}b)^2\\ \qquad B_1+2t(b-t_1)+2b_1l_1\end{array}\right.$

" CD $\left\{\dfrac{\tfrac{2b_1l_1{}^2}{12}+tb_1l_1(l+l_1)^2+\tfrac{bl^3}{12}-\tfrac{2}{3}(b-t_1)(\tfrac{1}{2}l-t)^2}{B_1+2t(b-t_1)+2b_1l_1}\right.$

$b^3+l t^2-t^3$
$12(b+l-t)$

principal axis, and set it oscillating. Count the number of oscillations in a minute, and substitute in the formula:

$$r^2 = \tfrac{1}{4}\left(\frac{140,796}{N^2}e - e^2\right)$$

in which N is the number of oscillations, and e is the distance from the point of suspension to the centre of gravity of the section, in inches. The cardboard should be heavy, otherwise it may be difficult to get it to continue to oscillate for so long a time as a minute. The oscillations in half a minute multiplied by two would of course be the same, but it is difficult to count them exactly, and any error is doubled in the result. For columns acting as braces, a similar process may be employed. However, their sections are not usually continuous; for example, the "Phœnix" column, No. 4 in the previous table, sometimes has spaces left between the sections which compose it, occupied by "thimbles" or iron tubes around the rivets. In cutting out the cardboard figure in such a case, the places occupied by the rivets should be represented merely by a narrow piece, sufficient only to hold the sections so that they shall oscillate together. The following are some other forms of bracing for sustaining compression.

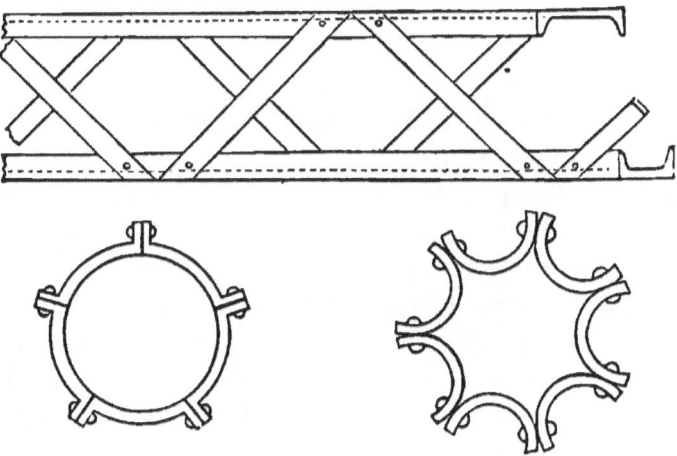

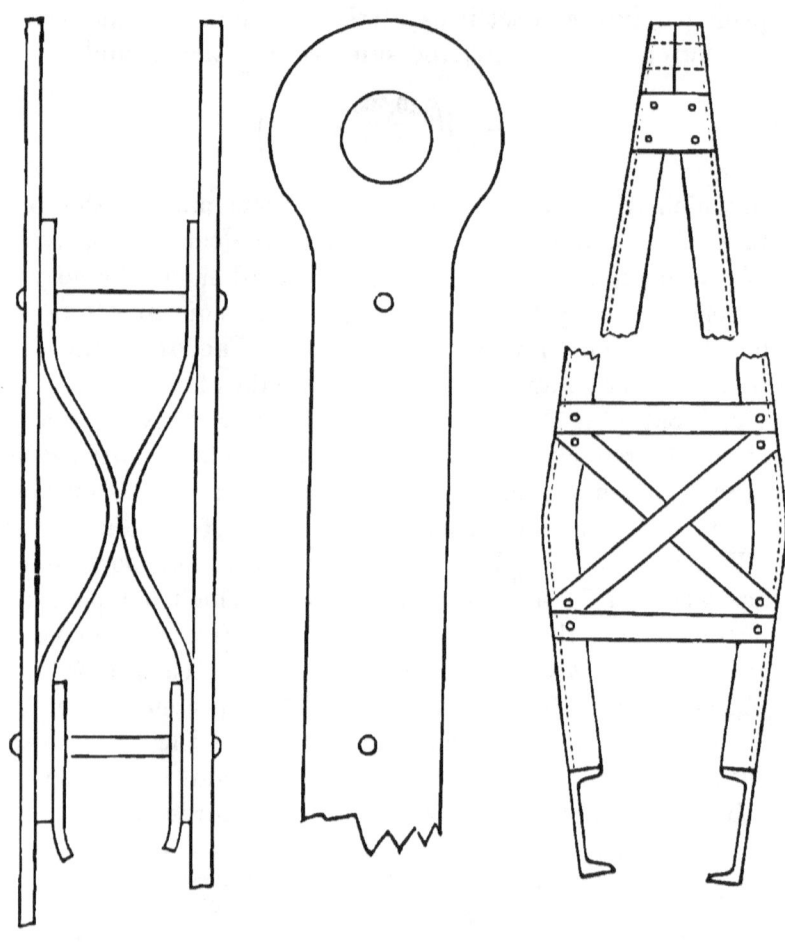

In pin connected bridges, it will often be necessary to rivet extra pieces to the chord around the pin holes, to distribute the stress over a large surface, so that the chord shall not fail by the pin crushing into it. Where a brace sustaining compression is composed of two channel irons, connected by pieces rivetted diagonally to the flanges, the least radius of gyration may be found by the same process with the card-board, substituting the narrow strips however, to represent the diagonal pieces, in the cross section.

The diagonal pieces may then be proportioned so that the column may be equally strong in both directions, to a transverse strain.

The following is the specification of Messrs. Wilson, Brothers & Co., of Philadelphia, for wrought iron bridges:

"WROUGHT IRON.

"All wrought iron must be tough, fibrous, uniform in quality throughout, free from flaws, blisters and injurious cracks, and must have a workmanlike finish. It must be capable of sustaining an ultimate stress of forty-six thousand (46,000) pounds per square inch on a full section of test piece, with an elastic limit of twenty-three thousand (23,000) pounds per square inch.

"All iron to be used in tension or subjected to transverse stress (except web plates) must have a minimum stretch of fifteen (15) per cent., under ultimate stress, measured on a length of eight (8) inches.

"All iron to be used in compression and for web plates of width not exceeding twenty-four (24) inches, must have a minimum stretch of ten (10) per cent. under ultimate stress, measured on a length of eight (8) inches.

"All iron for web-plates exceeding twenty-four (24) inches in width must have a minimum stretch of five (5) per cent. measured in length of eight (8) inches.

"All iron to be used in the tensile members of open trusses, laterals, pins, bolts, etc., must be double rolled after and directly from the muck bar (no scrap will be allowed) and must be capable of sustaining an ultimate stress of fifty thousand (50,000) pounds per square inch on a full section of test piece, with an elastic limit of twenty-five thousand (25,000) pounds per square inch and a minimum stretch of twenty per cent. in length of eight inches under ultimate stress.

"When tested to the breaking, if so required by the engineer, the links and rods must part through the body and not through the head or pin hole. Such tests must be at the expense of the contractor when the requirements of these specifications are not complied with.

"All wrought iron, if cut into testing strips one and a half (1½) inches in width, must be capable of resisting without signs of fracture, bending cold by blows of a hammer until the ends of the strip form a right angle with each other, the inner radius of the curve of bending being not more than twice the thickness of the piece tested. The hammering must be only on the extremities of the specimens, and never where the flexion is taking place. The bending must stop when the first crack appears.

"All tension tests are to be made on a standard test piece, of one and a-half (1½) inches in width, and from one-quarter (¼) to three-quarter (¾) inches in thickness, planed down on both edges equally, so as to reduce the width to one (1) inch for length of eight (8) inches. Whenever practicable, the two flat sides of the piece to be left as they come from the rolls. In all other cases, *both* sides of the test pieces are to be planed off. In making tests the stresses are to be applied regularly, at the rate of at least one (1) ton per square inch in fifteen seconds of time.

"All plates, angles, etc., which are to be bent in the manufacture must, in addition to the above requirements, be capable of bending sharply to a right angle, at a working heat, without showing any signs of fracture.

"All rivet iron must be tough and soft, and pieces of the full diameter of the rivet must be capable of bending until the sides are in close contact, without showing fracture.

"*Workmanship.*—All workmanship must be first class; all abutting surfaces must be planed or turned, so as to insure even bearings, taking light cuts so as not to injure the end fibres of the piece, and must be protected by white lead and tallow. Pieces where abutting must be brought into close and forcible contact by the use of clamps or other approved means before being rivetted together.

"Rivet holes must be carefully spaced and punched, and must in all cases be reamed to fit where they do not come truly and accurately opposite, without the aid of drift pins. Rivets must completely fill the holes and have full heads, and be countersunk when so required.

"All pin holes in pieces which are not adjustable for length must be accurately bored at right angles to the axis unless otherwise shown on the drawings, and no variation of more than one sixty-fourth of an inch will be allowed in the length between centres of pin holes. Pins must be carefully turned and no variation of more than one-thirty-second of an inch will be allowed between diameter of pin and pin hole. In the case where rough bolts are permitted, a variation of one-sixteenth of an inch will be allowed between diameter of bolt and hole. Thickening washers must be used whenever required to make the joints snug and tight.

"All iron must receive one (1) coat of boiled linseed oil before leaving works. All inaccessible surfaces are to be painted, preferably at the bridge site during erection, with one (1) heavy coat red oxide of iron in pure linseed oil. All iron to be scraped clean from scale before painting.

"GENERAL CONDITIONS.

"The whole of the construction to be first-class work, and in strict accordance with the drawings and these specifications. In the case of sub-contractors, the specifications are fully binding on them in every respect, and free access and information is to be given by them for thorough inspection of material and workmanship, and all required test pieces, etc., are to be provided as may be requested.

" In all cases figures are to be taken in preference to any measurements by scale.

" No alterations are to be made unless authorized by the engineers."

TRESTLE WORK.

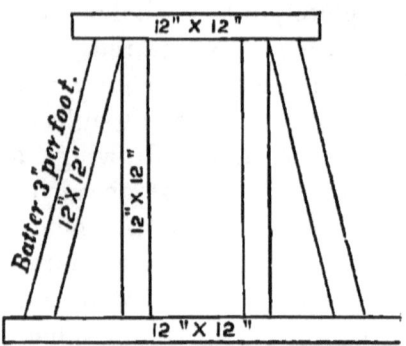

The above is the usual form. "The bents," or at least some of them, should be well braced together with 6 × 12 stuff, and placed about 10 feet apart. When more than 15 feet high, make it in two or more stories. The posts are generally mortised into the cap and sill pieces. In such cases if the mortise for the lower tenon is not cut entirely through, a hole should be bored to let the water out which will collect in it. Some engineers consider it better to cut no mortises or tenons at the lower end, but to make the connections with wooden dowel pins, two to each post, to prevent twisting. These should be large enough to fit tightly in a three-inch hole. The inclined posts should be cut off square, and rest in shoulders cut in the cap and sill. Iron straps may then be used to keep all secure.

When a mortise and tenon are used, the width of the tenon may be made from one-third to one-fourth the width of the stick. A tree-nail passes through both mortise and tenon, to hold them fast. It is usual to bore the holes for this, not directly opposite to one another, but that in the tenon a little nearer the shoulder than would be opposite that in the mortise. This will make the driving of the pin draw the post tightly against the cap. In a trestle designed by the writer for carrying water-pipe, a tenon 5 inches long on a 6 × 8 stick, in a cap of the same size, with a pin $1\frac{1}{2}$ inch in diameter, was made to draw $\frac{1}{8}$ of an inch, which proved to be a very little too much, as it cracked some of the caps; not enough, however, to destroy any or make them unfit for their duty. However, a draw of $\frac{3}{32}$ would have been better for this size of timber. One-eighth draw would do for 12 × 12 sticks.

High trestle-work is generally made of iron. It is usually made to batter in one direction, and it is better that the batter should be in a plane at right angles to the centre line of the bridge. Abroad it is often made to batter in a plane parallel to the centre line, but as the force due to the wind acting to overturn it is greater, when the structure is very high, than the reaction due to the force on the driving wheel, the additional strength due to the greater leverage obtained by spreading the base should be opposed to the wind pressure. The high piers built by American engineers usually batter at one and one-half inch to a foot, but the columns at opposite sides of the bridge are in vertical planes at right angles to it. The following formulas represent the strains in the several parts of such a pier in one of these vertical planes.

Let P = the pressure due to the wind on the train, and trusses on one pair of columns, found by multiplying the side surface in square feet of a train covering half of each adjacent span and the exposed surface of the half spans themselves by the maximum pressure of the wind on a square foot. Let P' = the pressure due to the wind

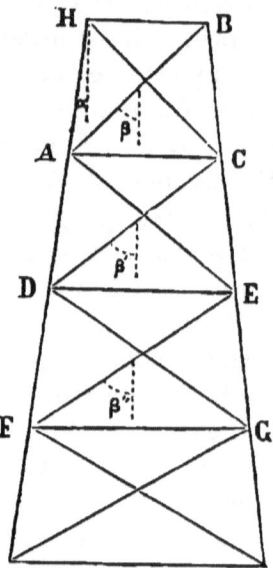

on one panel of the pier, on one pair of columns, found by multiplying one-half the surface exposed on the adjacent sides of the bent, by the maximum pressure of the wind.

Let W = the vertical load due to train and bridge on one column; for the outer columns it will be one-half the load on the adjacent spans of bridge and pier.

Let W^1 = the vertical load due to one panel of the pier on one column. Let α = the angle made by the columns with the vertical. The batter is usually, in American structures, 1½ inches to a foot, when α will = 7° 8'.

Let β, β', β'', etc., = the successive angles which the diagonal braces make with the vertical, from the top downwards.

Then,

Strain on $HB = P + W \tan. \alpha + \frac{1}{2} W' \tan. \alpha$ compression.

Strain on $AC = P \dfrac{\sin. (\beta - \alpha)}{\sin. (\beta + \alpha)} + P + W'' \tan. \alpha$ compression.

Strain on
$$DE = P\frac{\sin.(\beta-\alpha)\sin.(\beta'-\alpha)}{\sin.(\beta+\alpha)\sin.(\beta'+\alpha)\tan.\alpha} + P'\frac{\sin.(\beta'-\alpha)}{\sin.(\beta'+\alpha)} + P' + W'$$

Strain on
$$FG = P\frac{\sin.(\beta-\alpha)\sin.(\beta'-\alpha)\sin.(\beta''-\alpha)}{\sin.(\beta+\alpha)\sin.(\beta'+\alpha)\sin.(\beta''+\alpha)} + P'\frac{\sin.(\beta'-\alpha)}{\sin.(\beta'+\alpha)}\frac{\sin.(\beta''-\alpha)}{\sin.(\beta''+\alpha)} + P''\frac{\sin.(\beta''-\alpha)}{\sin.(\beta''+\alpha)} + P'' + W'\tan.\alpha$$

&c. = &c.

Strain on $AB = P\dfrac{\cos.\alpha}{\sin.(\beta+\alpha)}$ tension.

Strain on $DC = P\dfrac{\cos.\alpha\sin.(\beta-\alpha)}{\sin.(\beta+\alpha)\sin.(\beta'+\alpha)} + P'\dfrac{\cos.\alpha}{\sin.(\beta'+\alpha)}$ tension.

Strain on
$$FE = P\frac{\cos.\alpha\sin.(\beta-\alpha)\sin.(\beta'-\alpha)}{\sin.(\beta+\alpha)\sin.(\beta'+\alpha)\sin.(\beta''+\alpha)} + P''\frac{\cos.\alpha\sin.(\beta'-\alpha)}{\sin.(\beta'+\alpha)\sin.(\beta''+\alpha)} + P'\frac{\cos.\alpha}{\sin.(\beta''+\alpha)}\,\&c. = \&c.$$

Strain on $AH = P\dfrac{\cos.\beta}{\sin.(\beta+\alpha)} + \dfrac{W}{\cos.\alpha}$ compression.

Strain on
$$DA = P\frac{\cos.\beta}{\sin.(\beta+\alpha)} + P\frac{\cos.\beta\sin.(\beta-\alpha)}{\sin.(\beta+\alpha)\sin.(\beta'+\alpha)} + P'\frac{\cos.\beta}{\sin.(\beta'+\alpha)} + \frac{W}{\cos.\alpha} + \frac{W}{\cos.\alpha} \text{ compression.}$$

Strain on $FD = P\dfrac{\cos\beta}{\sin(\beta+\alpha)} + P\dfrac{\cos\beta\sin(\beta-\alpha)}{\sin(\beta+\alpha)\sin(\beta'+\alpha)}$
$+ P\dfrac{\cos\beta\sin(\beta-\alpha)\sin(\beta'-\alpha)}{\sin(\beta+\alpha)\sin(\beta'+\alpha)\sin(\beta''+\alpha)} + P'\dfrac{\sin(\beta'-\alpha)\cos\beta}{\sin(\beta''+\alpha)\sin(\beta'+\alpha)}$
$+ P'\dfrac{\cos\beta}{\sin(\beta''+\alpha)} + P''\dfrac{\cos\beta}{\sin(\beta''+\alpha)} + \dfrac{W+2W'}{\cos\alpha}$ compression.

Strain on AH = no tension.

Strain on $DA = \dfrac{W}{\cos \alpha} + \dfrac{W''}{\cos \alpha} - P\dfrac{\cos \beta}{\sin (\beta + \alpha)}$ tension, if a minus quantity; if it is a plus quantity, there is no tension on DA.

Strain on $FD = \dfrac{W'}{\cos \alpha} + \dfrac{2W''}{\cos \alpha} - P\dfrac{\cos \beta}{\sin (\beta + \alpha)} - P\dfrac{\cos \beta' \sin(\beta - \alpha)}{\sin(\beta + \alpha)\sin(\beta' + \alpha)}$

$- P' \dfrac{\cos \beta'}{\sin (\beta' + \alpha)}$ tension, if a minus quantity.

Make these calculations for the bridge loaded with its greatest load, and also unloaded; the pressure of the wind per square foot being taken the same in both cases. The tension on the lowest columns in the unloaded bridge (if any), due to the differences in the strains produced by the wind and load, may exceed that produced under the same circumstances with the loaded bridge.

Prof. Wm. H. Burr has called attention to the fact that the effect of the wind on the train is to raise the windward side and so increase the load on the leeward side, decreasing and increasing the loads on the respective columns as given above. Similarly the effect of the wind on the trusses when the top chord rests on the pier is to decrease the load on the leeward side, and increase it on the windward. The effect produced by the difference of these should therefore be subtracted or added (as the moment on the train or truss is the greater) to that produced by P above. This may be accomplished by writing $P \pm 2\, t \tan. \alpha$ for P and $W + t$ for W for the leeward, and $W - t$ for the windward column, in the formulas above, t being equal to

$$\dfrac{H\,h - H'\,h'}{l}$$

in which H is the pressure of the wind on the train covering the adjacent half spans, and h is the height of the centre of pressure above the point of connection of the truss with the piers, and H' and h' are similarly the total pressure of the wind on the half span and the depth of the

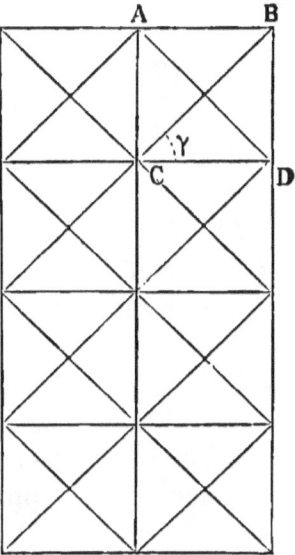

centre of pressure below the point of connection of the truss with the pier, and l is the distance HB in the figure.

In a direction parallel to the centre line of the bridge, the pressure on the braces is due to the action of the driving-wheel of the locomotive, and is found as follows:

Let $\gamma =$ the angle made by the brace BC with the horizontal. If T is the horizontal force exerted by the steam in the cylinder, and S is the stroke of the piston, and D is the diameter of the driving-wheel, the stress on AB, CD, &c., will be $T\frac{S}{D}$, and that on AD, &c., equal to $\frac{ST}{D\cos.\gamma}$

To find T, if the cylinder of the locomotive is d inches in diameter, and the pressure of the steam in the boiler 140 pounds per square inch,

$$T = 140\pi \frac{d^2}{4} = 109\, d^2$$

The value of $\frac{S}{D}$ is about two-fifths in the heaviest engines on the ordinary gauge.

The strain on AD resolved on AC must be added to the compression due to the weight of the train and bridge, and the effect of the wind, on the same piece.

The pressure in the steam cylinder is never as great as that in the boiler, and will likewise be less the faster the engine is running. If the engine, however, moves slowly from a state of rest the strains will approach those given above.

MEASUREMENTS OF BRIDGE SPANS.

As the young engineer may be called upon to stake out the masonry of bridge piers and abutments when there may be some difficulty in making a direct measurement at the place where the bridge is to be, descriptions will be given of the methods which have been employed in several actual cases of the kind, which presented some interesting problems.

First, the Mahoning Creek bridge, on the Allegheny Valley Railroad, in Pennsylvania.

This consisted of two spans each of 150 feet. Plugs were put in on the centre line at A and B, and a base line measured on the sandy shore AC, and the angle BAC likewise measured. The distance AB was then calculated, and the position of one of the abutments decided upon. The distance from this abutment to the centre of the pier was then, of course, known, and points D and E were assumed at known distances from the centre of the pier, in its centre line produced. The water was sufficiently shallow to allow long iron rods to be driven into the bed of the creek and project above the surface at low water. The angles $D A C$ and $D C A$ were calculated so that with a

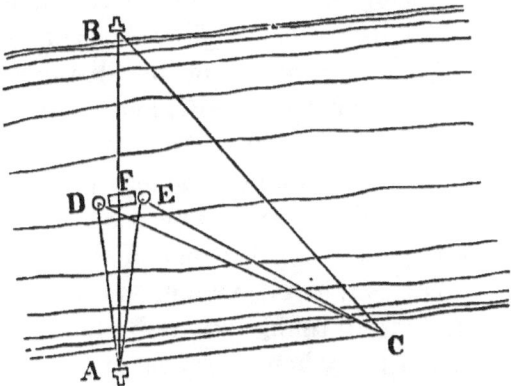

transit at both A and C with these angles turned off, the rodman, in a boat, could be directed one way or another until the iron rod was in line from both instruments; it was then driven down. Similarly a rod was driven at E. For calculating the angle $D A C$, the angles $D A F$ $\left(= \tan.^{-1} \dfrac{D F}{A F}\right)$ and $F A C$, (measured,) were added together. From $A D \left(= \sqrt{D F'^2 + A F'^2}\right)$ and $A C$, (measured,) and the included angle, calculated above, the angle $D C A$ was calculated. Similarly the angles to the point E were calculated. As the engineers' camp contained but one transit, the level was used in lieu of another. The transit having been set at A and the angle $D A C$ turned off, a plug was put in on the opposite side of the creek; the level was then set over the point A and sighted on the plug and clamped. The transit was then set at C

and the angle ACD turned off; by its means and the level the rod was set at D. This method, which was devised by the principal assistant engineer, though ingenious, was not perfectly satisfactory to the engineer in charge. The iron rods were used, however, for placing the wooden platform on which the masonry was built. Spikes were driven in its centre line at each end, and it was floated until they were at the proper distance from the rods. The masonry was then built, and settled the platform on the bottom. Just below the water-line an offset of six inches was left all round, the neat work beginning at that distance from the edge. The want of confidence in the method was principally because of the inferior character of the transit employed, which, to secure lightness, had a small graduated limb, and had seen some rough service. Accordingly, when the neat work was to be started, a stone was laid on the foundation to raise it out of the water, and a line was stretched from it to the abutment. The assistant then took his rodman in a boat, with a rod ten feet long, which was applied successively to the line, beginning at the abutment, a common pin being thrust through the line at the front end of the rod, to which the hind end was then held. This measurement was not in a straight line, because the string sagged. This was corrected by using inversely Weisbach's formula for the length of a suspension-bridge cable, having the span and "versin." $L = \frac{2}{3} \frac{V^2}{S}$; the versin. being estimated by holding a level rod at the middle and sighting from one end to the other, observing where the visual line cut the rod. This second measurement, by which the bridge was built, differed about one foot from the triangulated distance, which was allowed for in the framing of the bridge.

The next example is that of the Delaware River bridge,

on the North Pennsylvania Railroad route to New York. It was devised by Mr. D. McN. Stauffer, the engineer in charge

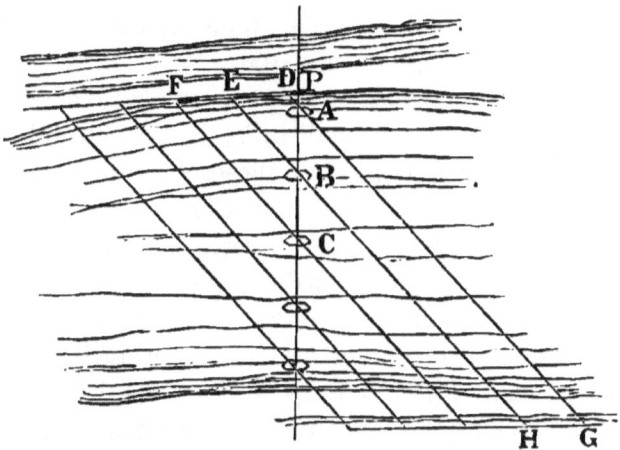

He took advantage of the cold winter weather, when the river was frozen over, and measured the spans carefully on the ice, marking the different centres of the piers A, B, C, etc. He then put iron plugs, one inch in diameter, four feet long, on the shore at D, E, F, etc., in line, and at equal distances apart. He set his transit over D, sighted on A, and put in a plug on the opposite side of the river at G; similarly H was put in on the line EB produced; etc. The piers were built on wooden platforms, which were floated into place when the ice had disappeared. The centre line having been drawn on each platform, and marked by a long pin on each side, and a short one with red flannel attached in the centre, a transit at P served to put it in proper line, while one at D, E, F, etc., successively served to correctly place the centres. An unforeseen difficulty arose from boys stealing some of the iron plugs, but the ground being frozen the holes remained, and they were easily replaced.

The final example is that pursued by Mr. L. L. Buck for

the Verrugas Viaduct, on the Callao, Lima and Oroya Railroad, in Peru. This viaduct is situated in some of the wildest scenery of the Cordilleras, at a height of nearly 6,000 feet above the sea. The slopes of the mountains are exceedingly steep, generally too much so to permit of ascending until a road is cut. There are often places, however, on the tops of small hills, or on their sides, where the slopes are less abrupt, and these were taken advantage of by the ancient inhabitants of the land for terracing in successive level steps, over which the water, which is necessary for irrigation, was conducted by canals. The climate is very dry, the rainy season being confined to six months, and even then consisting of merely light rains in the afternoon. There being no frost the hills retain the steep character of the volcanic disturbances which created them. The old Inca terraces, however, formed admirable places for measuring a base line.

The Verrugas viaduct consists of one span of 125 feet, and three spans of 100 feet, supported by abutments, and by three iron piers of heights of 177, 252 and 152 feet respectively, each pier measuring 50 feet across on the centre line of the railroad. The slopes were too abrupt on the sides to permit of descending on the centre line of the bridge.

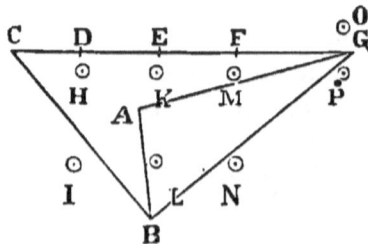

A base line *A B* was accordingly measured in the valley, 197 feet long, the positions of the piers being *D*, *E* and *F*. Points *C* and *G* were put in on the centre line near the faces of the abutments. The angles from A B to G were observed, and the distance B G calculated; from this distance, with observed angles at B and G, C G was calculated. Another base was then measured on a level terrace above, 231 feet long, but badly-situated with reference to C G. By a series of triangulations C G was again obtained. It differed from the first value by four-tenths of a foot. The first triangulation was supposed to be the most accurate, and stakes were put in from it at D, E and F, the centres of the piers, and guarded by turning off right angles, and putting plugs at H, I, K, L, M, N, O, P. As the bridges were made complete in the United States, and shipped to their destination, it was necessary that they should come together exactly on erection. Accordingly the following check was made on the above measurements and calculations: A line was measured on a level terrace, putting stakes every fifty feet, which were driven to the same level, and measured with a tape and spring balance. Stakes were also driven twenty feet apart, and the distance measured with a rod twenty feet six inches long, made deeper in the middle and tapering towards the ends, so as not to bend with its own weight. Marks were made on this rod exactly twenty feet apart. This latter measurement agreed with that of the tape and balance. Stakes were put in on this line at the points corresponding with the centre of each pier and the faces of the abutments. A long flat wire ("hoop-skirt wire") was then suspended over this base line, with one end fast and the other passing over a pulley, with a weight hung to the latter end. The centres of the piers and faces of the abutments were then "plumbed up" from the stakes, and marked, by winding, first with shoe-thread waxed, and then with white thread. The wire was afterwards suspended over the place which the bridge was to occupy, stretched with the same weight as before, and sights taken on it from the stakes at I, L,

N and *O*, using *H*, *K*, *M* and *P* as fore-sights. It was extremely satisfactory to Mr. Buck to find that one of the marks on the wire only varied one-half an inch, and all the others less than three-eighths of an inch, from the triangulated distances.

As a very interesting example of triangulation, the following, executed by Mr. O. F. Nichols, may be described:

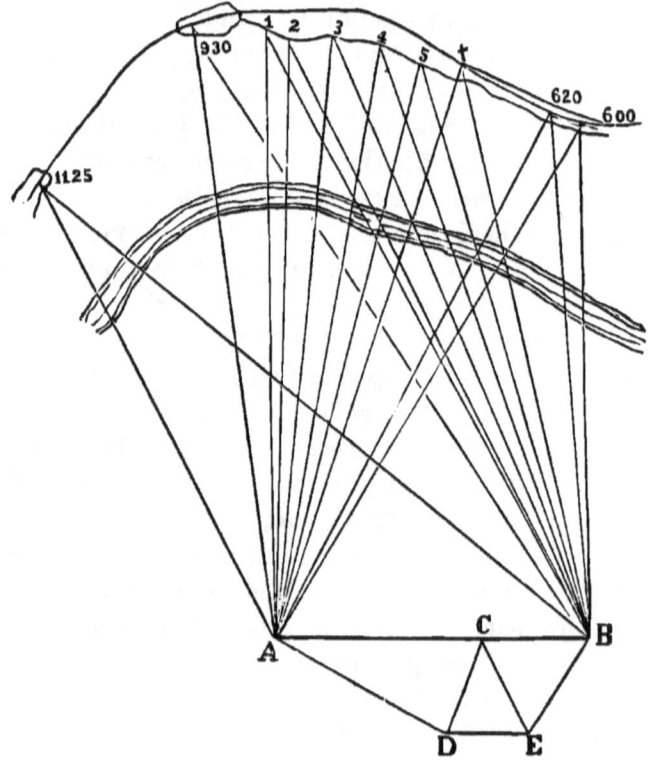

It was necessary to locate the Callao, Lima & Oroya Railroad over a very precipitous cliff, which was itself inaccessible. The line was run from each direction as far as possible, to stations 620 and 1125 in the figure. At 930 a point about grade could also be approached, and a small place was graded there, and fixed as a portion of the line.

The relative positions of these points were determined by triangulation from the opposite side of the valley. They were then plotted with the line as determined on each side of 1125 and 620, and connected on paper with proper lines, and tunnels driven to connect them. In locating these lines on paper, it was borne in mind that the tunnels should always continue in the rock, and not cross any seams that might run in from the face of the cliff. Points on this face were therefore likewise determined by triangulation and plotted. On the side of the valley from which the triangulations were made, there existed numerous Indian level terraces, made for purposes of irrigation, which were utilized for the base lines. The points A and B were fixed upon at convenient places, and an intermediate point C marked. A base line $D\ E$ was then measured very carefully several times. From it the distances $A\ C$ and $C\ B$ were calculated, and added together, giving the distance $A\ B$ as a longer base line. From A and B the successive points 620, 930, 1125, 1, 2, 3, 4, 5, †, etc., were observed; some of these consisting of discolorations of the rocks, produced by growing lichens, and others of marks made by firing rifle balls against the cliff near the grade line. Having afterwards calculated the various distances and angles, and plotted them, the line was fixed. From 930 to 1125 an interesting problem arose. The direction of the tangent at 930 with the lines to the extremities of the base line was fixed, 930 was taken as the beginning of the curve with a given radius, and the length of the curve was so calculated that a tangent at the other extremity should pass through 1125. The tunnel was driven from both ends and the headings met on the tangent. It was therefore necessary to run the curve to the tangent point very accurately, and the angle turned off should coincide with the centre line run from the other end. As an example of Mr. Nichols' accurate work under great difficulties, it may be mentioned that when the headings met there was an error of less than two inches, the tunnel being 537 feet long, the base line 886.79 feet long, and situated 1750 feet off, the work prosecuted under a tropical sun, in a country so malarious that the men died, during the sickly season, at an average of about thirteen a day. Between 930 and 3, similar calculations were made and a tunnel driven. In addition to the lines shown in the figure, two other base lines were calculated and measured as bases of verification; they are omitted, however, as they would too much complicate the figure.

MASONRY.

In masonry, the two important points required are to have good beds to the stones, and to have a good bond. When not closely watched, masons working for contractors will put "nigger-heads" into the wall, that is, stones from which the natural rounded surface of the stone or boulder has not been taken off. Each course should be made level before beginning the next one. There should be plenty of headers, to run into the wall as far as possible. A trick of the masons is to use "blind headers," or short stones that look like headers on the outside, but do not go deeper into the wall than the adjacent stretchers. When a course has been put on top they are completely covered up, and, if not suspected, the fraud will never be discovered unless the wall shows its weakness. The headers may taper off at the back in their width, but should retain their depth throughout. Rankine recommends that one-fourth the face of a wall should consist of headers, whose length should be from three to five times the depth of a course. Some engineers prefer masonry laid dry, as it requires a superior class of work, and imperfections are more easily detected. For railroads, however, it is apt to be shaken loose by the motions of the trains.

Mortar is made of a mixture of 1½ bushels of lime to 3 of sand, the lime to be slacked (one part water to one of lime) immediately before using, before the sand is put in; the sand to be sharp, and better if the grains are large (coarse). This amount will lay a perch of stone. Vicat says the best proportions are 2.4 of sand to 1 of lime (Rankine). If cement is used, not too much water should be mixed with it. It is found that the cement becomes hardest if just enough water is mixed to make it moderately dry, but capable of being finished and smoothed off with a trowel. The proportions are 1 of water to 4 of cement, by measure. Cement used neat is liable to crack. For this reason, and the sake of economy, it is mixed with an equal measure of sand, the larger and sharper the better. Above the water line, 2 parts of sand to 1 of cement may be used. For mixing cement and sand, an old wine barrel is better than a mortar box. In the latter it is difficult to prevent the sand from floating on top; that is, it is more difficult to get them thoroughly mixed.

It used to be the custom to lay about four inches from the face in cement mortar thick, and to grout the remainder with the same mortar, with its consistency so reduced by water that it would run into the interstices. As it is found, however, that this excess of water injures the quality of the cement, the practice is being given up.

FOUNDATIONS.

The best foundation is rock. Rankine says the crushing strain of limestone and sandstone is from 3,000 to 8,500 pounds per square inch, and granite 10,000 to 13,000. He says the actual pressure should in no case exceed one-eighth the crushing strain. He likewise says that foundations in gravel, hard clay, and sand are usually loaded with from 2,500 to 3,500 pounds per square foot. At Nantes, 17,000 on sand caused some settlement. The anchorage of the

Brooklyn Bridge is built on a wooden platform, three feet thick, which rests on sand 22 feet below the surface, with a pressure of four tons per square foot. (Trans. Amer. Soc. of Civil Engineers, Aug., 1874.) Mr. H. Leonard, in *Engineering*, Vol. 20, p. 103, says that, from the result of actual experiments in India, alluvial soil will safely bear 2,240 pounds to the square foot. He says the depth of such foundations should not be less than four nor more than eight feet. The offsets of the foundation should spread at an angle of 45 degrees, and no step should be less than 18 inches high; if less, it will break off. Sir Charles Fox says his experiments show that alluvial soil will bear 1,680 pounds per square foot. Rankine says the usual rule for spreading the foundation of a wall is to make the breadth of the base $1\frac{1}{2}$ times the thickness of the body of the wall in compact gravel, and twice the thickness in sand and stiff clay.

The foundation, especially in sand and alluvial soil, should be kept from the effects of running water. This may sometimes be avoided by "rip-rapping" round the foundation.

A common way of building on a gravel foundation in streams where the water does not exceed five feet in depth, is to build on two courses of 12 × 12 inches squared timber, laid close, at right angles to each other, and spiked at each intersection. This is floated in place, and the masonry built until it sinks. For deeper water, there would be a risk of upsetting before reaching the bottom, and guide piles must be driven on the outside. Foundations on gravel or rock, in water, are often built on cribs. A framework, the size of the crib, should be floated to the proper place, and soundings taken all round it at intervals of about three feet, with some intermediate ones across. The bottom of the crib is then shaped to the surface given by the soundings, and so will have an even bearing. The crib is made in open cells, about 6 feet square, with a floor at the bottom to hold the loose stones which are filled in to sink it. The outside is made with 12 × 12 squared sticks, fitting close, laid horizontally. Where the sticks forming the cells come through the sides, they are dovetailed to them, and spiked besides. All these side sticks are spiked together with one inch spikes which are long enough to go through three of them, so that each one is spiked to the two below.

The sides of the crib are generally made with a batter of 1½ inch to a foot.

When the bottom is a soft mud, piles must be resorted to for a foundation.

PILE DRIVING.

Weisbach's formula:

$$L = W\left(\frac{W}{W+W'}\right)\frac{h}{d}$$

where W = the weight of the ram in pounds.
W' = the weight of the pile in pounds.
d = the depth which the last stroke drives the pile in inches.
h = the height of fall in inches.
L = the load which the pile will just bear in pounds.

The Dutch engineers use a similar formula, except that they use for d the average penetration per blow, got by taking the whole penetration in, say, 100 blows and dividing by 100. They also use a factor of safety of ¼.

If W' is supposed to be so small in comparison with W that it may be neglected, and a factor of safety of ⅛ is taken, we have Sanders' formula :

$$L = \frac{Wh}{8\,d}$$

where L is the safe load the pile will bear.

Rankine's formula, supposing the pile to be sustained by the friction on the sides, is

$$L' = \sqrt{\frac{4\,E\,S\,W\,h}{l} + \frac{4\,E^2\,S^2\,d^2}{l^2}} - \frac{2\,E\,S\,d}{l}$$

where E = the modulus of elasticity of the pile.
S = the sectional area of the pile in inches.
l = the length of the pile in inches.

We may take $E = 1{,}440{,}000$.
$S = 1\frac{1}{2}$ square feet = 216 square inches for an average.
$l = 30$ feet = 360 inches for an average.

Then the formula reduces to

$$L' = 1{,}859 \sqrt{W\,h + 864{,}000\,d^2} - 1{,}728{,}000\,d.$$

Rankine recommends a factor of safety of 5, which will reduce the equation to

$$L = 372 \sqrt{W\,h + 864{,}000\,d^2} - 345{,}600\,d.$$

Trautwine gives (reducing the value of his letters to the same measure):

$$L' = \frac{27\,W\sqrt[3]{h}}{1 + d} \text{ for the extreme load, and}$$

$$L = \frac{9\,W\sqrt[3]{h}}{1 + d} \text{ for the safe load.}$$

Rankine states that, according to the best authorities, the piles should be driven until $d = \frac{1}{150}$ of an inch, while Trautwine says the French consider it sufficient for d to equal $\frac{1}{75}$ of an inch, d being obtained in each case by dividing the total penetration in the last thirty blows by thirty. Many engineers, however, are satisfied to specify that the last blow with a 2,000 pounds hammer, falling 30 feet, shall not drive the pile more than half an inch. Rankine's formula takes a simpler form when h is expressed in feet. It then bears some resemblance to the formula given by McAlpine; the latter, however, being evidently derived from wrong hypotheses.

Rankine's formula, when h is in feet, and $d = \frac{1}{150}$, becomes

$$L' = 80 \left\{ 80\sqrt{Wh} - 144 \right\} \text{ for ultimate load, and}$$

$$L = 16 \left\{ 80\sqrt{Wh} - 144 \right\} \text{ for safe load.}$$

If W is expressed in tons, an approximate formula is

$$L' = 135 \sqrt{Wh} \text{ for ultimate load, and}$$

$$L = 27 \sqrt{Wh} \text{ for safe load.}$$

Rankine gives, as the limit of safe load on piles which reach firm ground, 1,000 pounds per square inch of head, and for the safe load on those which rely only on the friction of the mud against the sides, 200 pounds per square inch of head. Some actual experiments by Mr. E. T. D. Myers, on piles resting on a liquid mud, driven 45 feet deep, made nineteen hours after driving, gave 62 pounds only as the bearing power per square inch of head, this resistance, however, increasing with time. Some experiments in England (Van Nostrand's Magazine, Vol. 25, p. 275,) give 440 pounds per square inch of head as the withdrawing power of piles, or 1,875 pounds per square foot of contact for piles in stiff blue clay.

Rankine says the diameter should never be less than $\frac{1}{20}$ of the length. This probably refers only to piles resting on a hard stratum. Some of the piles supporting the bridges in the Hackensack marshes are 70 feet long, but *not* 3½ feet in diameter. The piles of the bridges which carry the Philadelphia, Wilmington & Baltimore Railroad over the Gunpowder River never reached a solid bottom. They are very long, however, and have proved perfectly satisfactory. Rankine says the best material is elm, and they should be driven butt downward. The ends should be sharpened to a point, whose length is 1½ times the diameter. The piles of the South Street Bridge, Philadelphia, were not sharpened, but were cut off square, to increase the bearing surface.

Piles should not be less than ten inches in diameter at the small end.

The bearing power of discs on iron piles in sand is five tons per square foot, according to Brunlees.

For driving piles through boulders and gravel, a heavy ram and small fall is the best. For example, a ram of 50 cwt. and a fall of 8 to 10 feet. For driving through sand, the blows should be delivered rapidly, so that the sand should not have time to compact itself about the pile in the intervals. A gunpowder pile driver is good for this purpose. McAlpine states that the surface friction in driving cast-iron cylinders 6 feet in diameter, through rocky gravel, was one-half a ton per square foot. General Smith found the friction of sand on an iron cylinder only $1\frac{39}{100}$ pounds per square inch. Gaudard gives the friction between cast-iron cylinders and gravel at 2 to 3 tons per square yard for small depths, and 4 to 5 tons per square yard for depths of 20 to 30 feet. He also says that piles at La Rochelle, in soft clay, can support 164 pounds per square foot of lateral contact, and at Lorient, in silt, 123 pounds. According to some observations made by Mr. Stauffer, at the South Street Bridge, in Philadelphia, the frictional resistance of mud on cast-iron cylinders was only 46½ pounds per square foot of surface.

For driving piles in sand, a great deal of labor can be saved by applying a water jet to loosen the sand. With hollow iron piles, the water may be forced down the interior of the pile, and its own weight will carry it down rapidly. With wooden piles, a notch is made at the end of the pile, in which the nozzle of a 1½ inch diameter hose is inserted. Water is then forced through this nozzle.

When masonry is supported by piles, care should be taken that no horizontal thrust comes upon it, unless the piles are specially braced for it. On the Delaware Extension of the Pennsylvania Railroad there is a mass of masonry 38 feet long, 12 feet wide, and 24 feet high, which serves as the abutment to the bridge over the Schuylkill River. When, many years ago, the embankment was built behind it to about half its height, the abutment began to move outward. The embankment then had to be removed and replaced with trestles. The retaining walls of the Chestnut Street Bridge, in Philadelphia, were built on platforms resting on piles, a separate platform for each wall, the walls running back parallel to the centre line of the street. These platforms were not tied together. Some time after the completion of the work the walls began to show signs of being pushed over. Buttresses have been built to prevent their further movement, but it is feared that it still continues, though certainly very slowly. If the platforms had been tied together at first by balks of timber, or otherwise, the thrusts of the walls would have been neutralized and their failure prevented.

The piles on which the "bulkheads" of the new docks at New York rest are braced together so as to prevent their being pushed out by the thrust of the material that they sustain behind them.

TRACK-LAYING

After the road is brought to sub-grade, the centre line is re-run, and stakes are set on each side of the road-bed at $4\frac{1}{2}$ feet off on a single-track road, and 100 feet apart, except on curves sharper than 3 degrees, where they should be 50 feet apart. These stakes are put one foot above the sub-grade, and give the top of the ballast. On curves, the outer one is elevated $\frac{1}{10}$ of a foot for each degree of curve above the inner one, which carries the grade. This gives the elevation of the outer rail $\frac{1}{2}$ an inch for each degree of curve, which is what it ought to be, supposing a speed of 27 miles an hour. The rule for track-layers to have when there are no stakes set, is, the elevation is equal to the middle ordinate of a chord of 48 feet of the curve.

The track-layer places a wooden straight-edge, 8 inches wide, on the stakes at two consecutive stations, and has two pieces of wood, 8 inches long, held upright on a tie at the places where the rails come. The tie is then driven down until the visual plane across the straight-edges just touches the tops of the blocks. Having set three intermediate ties in this way, the remaining ones are set with a straight-edge 15 feet long, laid on two of the ties already set. All the ties having been set, the half-gauge is measured off at the stakes, and the rail spiked fast, the portion between two stakes being lined by eye. One line of rails having been spiked, the other is spiked with a gauge-rod applied at every tie.

For bending rails to a curve, they are allowed to fall on two supports, placed at some distance apart, and the stored-up energy due to gravity produces the required result. (See *The Engineer* for Aug. 24, 1877, p. 139.)

A fall of 2 feet 2 inches on supports 18 feet apart, gives a curve of $\frac{3}{16}$ of an inch, corresponding to a curve of 2,970 feet radius; similarly,

A fall of 2 feet 8 inches gives a ver. sin. of $\frac{2}{3}$ of an inch, corresponding to a radius of 1,980 feet.

And a fall of 3 feet gives a ver. sin. of $\frac{9}{16}$ of an inch, corresponding to a radius of 990 feet.

For sharper curves a rail-bending machine must be employed.

In giving the elevation to the outer rail, it, of course, has to be given gradually, and it is the usual custom to begin back of the point of curve a sufficient distance to procure any permissable maximum grade to the rails which lead to the outer side of the curve. (Mr. Froude, quoted by Rankine, says this grade should not be more than one in three hundred. Mr. Geo. W. Kittredge, in *Engineering News* for June 18, 1881, says that one inch in sixty feet, or one in seven hundred and twenty, is found by experience on the Pittsburgh, Cincinnati & St. Louis Railroad, to be as great a grade as is desirable.) If, however, the curve is shifted inwards a certain amount and a new curve begins at some point on it, tangent to it, with a common radius of curvature at that point, but with radii continually becoming longer until it reaches the tangent with an infinite radius at the old point of curve; the elevation can be suited at each point to the curvature at the same point, and the condition that the elevation is always inversely proportional to the radius of curvature, is satisfied. Such a curve is called a "curve of adjustment," and its proper form is stated by Rankine to be an "elastic curve," which possesses the property of starting at a point tangent to a line, and having its radius of curvature at any other point inversely proportional to its distance from the tangent point. The "hydrostatic arch" is such an elastic curve, and could be approximated by a three-centre curve similar to the approximate geostatic arch on a previous page. It is simpler and better, however, to take another curve, which likewise approximates to the elastic curve. Prof. Rankine, in his Civil Engineering, Art. 434, gives the equation of this curve, which may be put in this form :

$$y = \frac{x^3}{6\, e\, R\, k}$$

(this differs from a cubical parabola, because x is measured on the curve, while in the cubical parabola it is a rectilinear co-ordinate) in which x is the distance measured on the circular arc, from the new point of compound curvature to a point whose ordinate from the original shifted curve is y; R is the radius of the circular curve, e is its proper elevation, and k is the reciprocal of the grade which it is determined to give the outer rail to obtain the elevation recommended above as 300. The distance of the point of compound curvature from the original point of curve is $\frac{k\,e}{2}$ and the amount that the circular curve is to be shifted over is $\frac{k^2\,e^2}{6\,R}$. If k is taken at 300 and $Re = \frac{5730}{24}$, the equation of the curve becomes

$$y = \frac{x^3}{429750}$$

the amount for the curve to be shifted over becomes

$$\frac{855023437\tfrac{1}{2}}{R^3}$$

and the length of the curve of adjustment becomes

$$\frac{35812\tfrac{1}{2}}{R}$$

(If $k = 720$, as recommended by Mr. Kittredge, the equation becomes

$$y = \frac{x^3}{1031400}$$

the amount for the curve to be shifted over becomes

$$\frac{4924935000}{R^3}$$

and the length of the curve of adjustment becomes

$$\frac{85950}{R})$$

To apply these equations, when the line is being staked out for the track, the tangents are run in on the original located line; a distance is measured along the original curve from the point of curve and point of tangent equal to the length of the curve of adjustment given above, and stakes are offset from these points on the inside of the curve to the distance the curve is to be shifted over, also given above; between these offset stakes the original curve is to be run in. Between the offset stakes and the point of curve and point of tangent, a curve is to be run in by calculating as many offsets from the circular curve as may be deemed necessary, by means of the equation of the curve of adjustment. One such intermediate stake will usually be sufficient for even very sharp curves. For a curve of 3 degrees, it is seen that the distance the curve is to be moved over is, by the equation, only about one and a half inch; for curves less sharp than this, we need scarcely go to the trouble of putting in a curve of adjustment, but use the original located line, and run out the elevation of the outer rail on the tangent. (If, however, Mr. Kittredge's value of e be used, a 3-degree curve should be shifted about $8\frac{1}{2}$ inches.) At points of compound curvature, a curve of adjustment should likewise be put in, if there is a very great difference in the radii. It is not often, however, that a sufficient difference exists to make it worth while to calculate the offsets, in the neighborhood of the point of compound curvature, from the shifted curves.

Some engineers, especially in Europe, widen the gauge on sharp curves, in order to lessen the resistance to the motion of the cars. This resistance will be directly as the rigid wheel base, and inversely as the radius of the curve. The additional width of gauge should be proportional to this resistance. A form of equation may then be written for the additional width,

$$w = \frac{a}{R} - b$$

in which a is a constant depending on the rigid wheel base, R is the radius of the curve, and b is a constant depending on the play of the wheels in the gauge.

The practice on the Bavarian railroads in Europe seems to correspond to a value for a of 1,115. Mr. Bouscaren widened on the Cincinnati Southern Railroad by a different rule. It seems to correspond, however, to a value of a of 430. In both cases b may be taken at $\frac{1}{10}$. The formula would then become for American rolling stock,

$$w = \frac{430}{R} - .1$$

R being in feet, and w in inches.

The writer is perfectly aware of the shortcomings of this formula, and only offers it as a substitute until reliable experiments give a firm foundation on which to build a better one. Mr. Doane believes in widening on American roads as much as on European ones. (See *Engineering News*, Vol. 8, p. 76.) It should be observed that the additional width should not in any event be more than one and one-half inch with wheels of the ordinary tread. This width, by the formula, would correspond to a curve of 269 feet radius. The rigid wheel base of European cars is 12 to 15 feet, and that of American from 5 to 9 feet. That of the largest locomotives is about 16 feet in Europe, and about 13 in the United States.

In Prussia the gauge is only widened on curves of less than 1,000 feet radius, and never more than one inch on the 4 feet 8½ inches gauge. (Molesworth's Pocket Book.) It will be observed that no widening is required by the formula when the radius is 4,300 feet or over.

The following is the cross-section of the Richmond, Fredericksburg & Potomac Railroad, designed by Mr. E. T. D. Myers, which is recommended for its excellence:

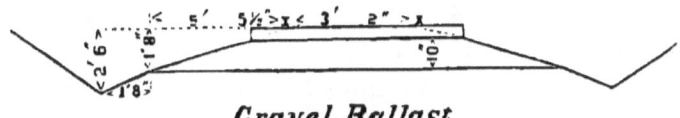

Gravel Ballast.

Below is likewise given the cross-section and specifications for the Pennsylvania Railroad, as prepared by Mr. W. H. Brown:

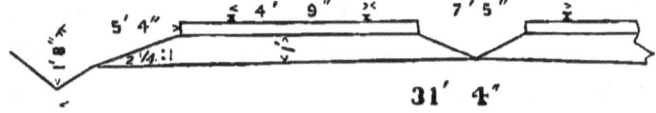

This is for gravel ballast; broken stone ballast is piled up level with the top of the tie, both between the tracks and at the ends.

" SPECIFICATIONS FOR A PERFECT SUB-DIVISION—PENNSYLVANIA RAILROAD.

Superstructure.

" 1. The track must be in good surface; on straight lines the rails must be on the same level, and on curves the proper elevation must be given to the outer rail, and carried uniformly around the curve. This elevation should be commenced from 100 to 150 feet back of the point of curvature, depending on the sharpness of the curve, and increased uniformly to the latter point, where the full elevation is attained. The same method should be adopted on leaving the curve.

"2. The track must be in good line.

"3. The splices must be properly put on, with the full number of bolts, nuts, stop-washers and stop-chairs. The nuts must be screwed up tight.

"4. The joints of rails must be exactly midway between the joint ties, and the joint on one line of rails must be opposite the centre of the rail on the other line of the same track. In winter a distance of five-sixteenths of an inch, and in summer one-sixteenth of an inch, must be left between the ends of the rails, to allow for expansion.

"5. The rails must be spiked both on the inside and outside on each tie, on straight lines as well as on curves.

"6. The cross-ties must be properly and evenly spaced, sixteen ties to a 30-feet rail, with 10 inches between the edges of bearing surfaces at joints, with intermediate ties evenly spaced a distance of not over 2 feet from centre to centre, and the ends on the outside, on double track, and on the right-hand side going north or west on single track, must be lined up parallel with the rails.

"7. The ties must not, under any circumstances, be notched, but should they be twisted, must be made true with the adze, and the rails must have an even bearing over the surface of the ties.

"8. The switches and frogs must be kept well lined up, and in good order. Switches must work easily, and safety blocks must be attached to every switch head.

"9. The switch signals must be kept bright and in good order.

Road-bed and Ballast.

" 10. The ballast must be broken evenly and not larger than a cube that will pass through a 2¼ inch ring. There must be a uniform depth of at least 12 inches of clean broken stone under the ties. The ballast must be filled up evenly between, but not above the top of the ties, and also between the main tracks and sidings, where there are any. In filling up between the tracks, coarse, large stones must be placed on the bottom, in order to provide for drainage, but care should be taken to keep the coarse stone away from the ends of the ties. At the outer end of the ties the ballast must be sloped off evenly to the sub-grade.

" 11. The road crossing planks must be securely spiked, the planking should be three-quarters of an inch below the top of the rail, and 2½ inches from the gauge line. The ends and inside edges of planks should be beveled off.

Ditches.

" 12. The cross-section of ditches at the highest point must be of the width and depth as shown on the standard drawing, and graded parallel with the track, so as to pass water freely during heavy rains and thoroughly drain the road-bed.

" 13. The lines must be made parallel with the rails, and well and neatly defined.

" 14. The necessary cross drains must be put in at proper intervals.

" 15. Earth taken from ditches or elsewhere must be dumped over the banks and not left at or near the ends of the ties, but distributed over the slopes. Earth taken out of the ditches must not be thrown on the slope [of a cut].

" 16. The channels of streams for a considerable distance

above the road should be examined, and brush, drift and other obstructions removed. Ditches, culverts, and box drains should be cleared of all obstructions, and the outlets and inlets of the same kept open to allow a free flow of water at all times.

Policing.

" 17. The telegraph poles must be kept in proper position, and trees near the telegraph line must be kept trimmed to prevent the branches touching the wires during high winds.

" 18. All old material, such as old ties, old rails, chairs, car material, etc., must be gathered up at least once a week and neatly piled at proper points.

" 19. Briers and undergrowth on the right of way must be kept cut close to the ground.

" 20. Station platforms and the grounds about stations must be kept clean and in good order."

Many engineers prefer placing the joints opposite each other, and if the track is kept well surfaced, it is probably better. It has the following advantage : The track should be of equal elasticity throughout. The joints are rarely of the same stiffness as the solid rail ; by placing the joint-ties nearer together, however, the stiffness of the track at the joints, when they are opposite, may be the same as at the other ties, whereas if the joints alternate, placing the ties nearer together at a joint makes the line of rails at the point opposite more rigid than at the adjacent parts. These remarks apply only to suspended joints. These were adopted because it was found that the rail-heads were less battered at the ends with them than when the joints rested on a tie. It is now said, however, that steel rails do not become battered like iron ones when they rest on a tie. If this is true, and suspended joints are abandoned, it seems that the method of breaking joints with the rails is the true one.

For information in regard to organizing gangs of track-layers in unsettled countries, with a description of the boarding cars for the men, etc., see *Engineering News*, Jan. 21st, and April 22d, 1882.

SWITCHES.

To find the distance from the point of a switch to the point of the frog on a curve:

The switch may curve to the outside or the inside of the curve.

The formulas for the distance are the same.

$$tan. \tfrac{1}{2} \alpha = \frac{g}{2 R \, tan. \tfrac{1}{2} F} = \frac{g \, n}{R}$$ (where n = the number of the frog).

$$x = \frac{g \cos. \tfrac{1}{2} \alpha}{sin. \tfrac{1}{2} F}$$

where R is the radius of the curve, g is the gauge, and F is the angle of the frog.

The value of the radius of the turnout differs in the two cases as below.

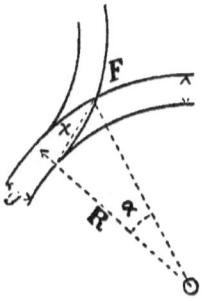

$$R = \frac{\tfrac{1}{2} x}{sin. \tfrac{1}{2} (F - \alpha)} - \frac{g}{2}$$

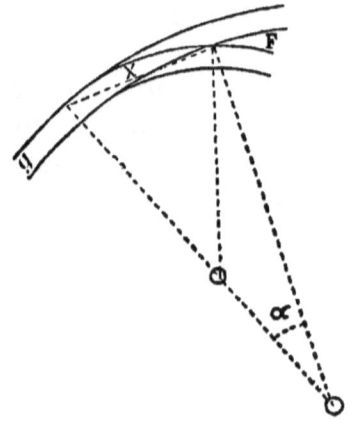

Radius of turnout $= \dfrac{\frac{1}{2}x}{\sin.\frac{1}{2}(F \pm \alpha)} - \dfrac{g}{2}$ { upper signs for inside.
lower signs for outside.

To find the distance from the point of switch to point of frog on a tangent:

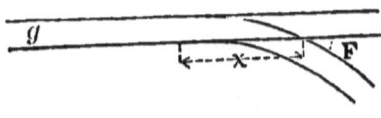

$$x = g\,\dfrac{1 + \cos. F}{\sin. F} = \dfrac{g}{\tan.\frac{1}{2}F}$$

It is usual to designate frogs by numbers which express the relation between the base and altitude of the triangle forming the point of the frog.

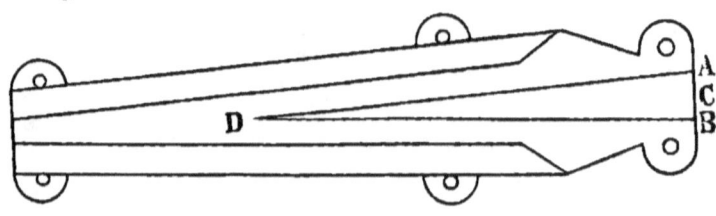

Thus, a No. 8 frog is one whose length is 8 times the base; $DC = 8 AB$.

The above equation then reduces to
$$x = 2gn,$$
where n = the number of the frog.

The radius of the turnout is $= \dfrac{x}{\sin. F} - \tfrac{1}{2} g$.

For a No. 8 frog, $F = 7° 9\tfrac{1}{4}'$. It is the usual practice to spike down 5 feet on each side of the frog straight, or, calling the distance from point to end of frog 2 feet, there would be 12 feet straight. The g in the formula (for a 4 feet 9 inches gauge) would be reduced to 3 feet $10\tfrac{1}{2}$ inches. In this case, then, $x = 62$ feet and $R = 495.72$, or about an 11° 35' curve. Five feet of the switch rail is spiked fast. In order to have a throw of $5\tfrac{1}{2}$ inches, the switch rail should be 27 feet long. The distance from the movable end of the switch rail, or point of switch, to the point of the frog is, then, 47 feet. This is an ordinary switch on railroads.

To find the distance from frog to frog on a crossing:

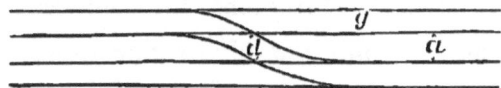

Call the distance from point of frog to point of tangent $= c$.

The distance measured on the track from frog to frog:
$$e = \dfrac{a}{\tan. F} - \dfrac{g}{\sin. F} - c \cos. F.$$

Then,
$$d = \sqrt{e^2 + a^2}$$

For a No. 8 frog, when $g = 4\tfrac{3}{4}$, $a = 7\tfrac{1}{4}$, and $c = 7$,
$$e = 12.67,$$
$$d = 14.60.$$

To find the proper angle for the middle frog in a three-throw switch:

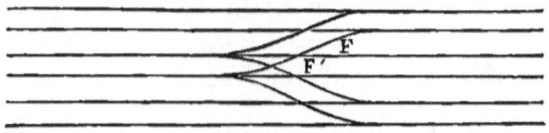

$$Ver.\ sin.\ \tfrac{1}{2}\ F' = sin.^2\ \tfrac{1}{2}\ F.$$

When two No. 10 frogs are used, we find $F' = 8° 6'$, and the third frog should be a No. 7. When two No. 8 frogs are used, the other should be a No. 5.64, or, say 6.

Bill of crossing lumber, single switch:

    3 ties.................... 9½ feet long.
    2 " ....10 " "
    2 " ....10½ " "
    2 " ....11 " "
    2 " ....11½ " "
    1 " ....12 " "
    5 " ....14 " "
    2 " ....15 " "
    1 " ....16 " "
    1 switch tie 15 feet long, 16 inches wide.

Bill of crossing lumber, double connection:

    6 ties.................... 9½ feet long.
    2 " ....10 " "
    4 " ....10½ " "
    4 " ....11 " "
    4 " ....11½ " "
    2 " ....12 " "
    16 " ....19 " "
    2 switch ties 15 feet long, 16 inches wide.

Bill of crossing lumber, triple connection:

    2 ties.................... 9 feet long.
    3 " ....10 " "
    2 " ....11 " "
    2 " ....12 " "
    3 " ....13 " "

1	"	14	feet long.
2	"	15	" "
1	"	16	" "
3	"	17	" "
3	"	18	" "
4	"	19	" "
2	"	20	" "

1 switch tie 15 feet long, 16 inches wide.

CROSS TIES.

These should be hewed (not sawed nor split) on two sides, cut square at the ends, and stripped of the bark before delivery. They should be 8½ feet long and 8 inches thick. Three-fourths of them should measure not less than 8 inches across the hewed surfaces, and one-fourth not less than 10 inches. They should be piled in square piles of about 50 each, the ties crossing each other at right angles in alternate layers. Each pile should be separated from the rest, so that a man can pass around each one to inspect the ties.

Public road crossings at grade :

The space between the tracks is covered with plank, 3⅜ × 8 inches, 16 feet long, spiked to the ties, and leaving 4 inches clear by the rail for the wheel flanges. Planks are also spiked to the ties on the outside of each rail.

RAILS.

A width of 4 inches is sufficient to prevent the rails from cutting into oak ties, and 4¼ inches for chestnut ties, when not spaced more than 2½ feet apart. If the base is made more than this, the difficulty of bending the rails to a curve becomes an objection. The stem of an iron rail need not be more than ½ an inch thick, nor that of a steel rail more than $\frac{7}{16}$ of an inch. 60-pound rails are made 4½ and 4¼ inches high, 50-pounds 4 inches, and those under 50-pounds 3½ inches.

For further information and sections of rails, reference may be made to the Reports in the Transactions of the American Society of Civil Engineers, Vol. 3, p. 106, and Vol. 4, p. 136, as well as to the *Railroad Gazette* for November 27th, 1875, and March 11th, 1881. Mr. Welch, in the second of the Reports, gives the following as the rules to be observed in the manufacture of iron rails:

"Select the stock best adapted to each part, the hardest metal for the head, the strongest for the base. Use only gray metal, not white. Put no old rails into the head or base; puddle thoroughly, or the metal will not weld thoroughly. Cut off and throw out all ends of puddle bars; make the top slab about $1\frac{1}{4}$ inch thick; thicker will not heat before the bars burn; pile 8 inches square. The top slab should be four rolled (thrice heated), the bottom thrice rolled, and the stem twice rolled; no puddle bars in the rail. Each heating should be uniform and thorough, without burning, or the welding will be imperfect."

The following are the specifications for steel rails adopted on the Pennsylvania Railroad by Mr. W. H. Brown:

"As it is the desire of the Pennsylvania Railroad Company to have on the roads under their control none but first-class tracks in every respect, and as the rails laid down on these tracks form an important part in the achievement of this result, the Pennsylvania Railroad Company have found it necessary to make certain demands in regard to the manufacture of their steel rails, with which the different rolling mills and rail inspectors will be required to comply.

"1. That the steel for rails shall be made in accordance with the 'pneumatic' process, and contain not less than $\frac{30}{100}$ nor more than $\frac{50}{100}$ of 1 per cent. of carbon.

"2. That a test bar, 1 foot long and $\frac{3}{4}$ of an inch square, is to be taken from the head of a rail made from each charge. This bar is to be supported at each end, leaving a space of 8 inches between bearings, and, by a sudden pressure on the centre of the bar, deflected to an angle of 80° from a straight line. This angle will be made the limit between good and bad tests. If the bar bends 80° without breaking, the charge will be considered good; if it breaks before bending, or shows signs of fracture at this angle, two more bars are to be tested; if both these bars stand the test, the charge will be considered first-class steel; but if one of them breaks the charge will be rejected and called second-class steel.

"3. The result of the test of each charge of which the Pennsylvania Railroad Company is to receive rails, and of which an official record is kept at each mill, is to be exhibited to the rail inspector.

"4. That the number of the charge and place and year of manufacture shall be marked in plain figures and letters on the side of the web of each rail.

"5. That the sections of the rails rolled shall correspond with the respective templates issued by the Pennsylvania Railroad Company, showing the shape and dimensions of the different rails adopted as their standard.

"6. That the space between the web of the rail and template, representing the splice-bar, shall not be less than $\frac{1}{4}$ of an inch, nor more than $\frac{3}{8}$ of an inch.

"7. That the weight of rails shall be kept as near to standard weights as can be demanded after complying with Section No. 5.

"8. That circular holes, 1 inch in diameter, shall be drilled through the web in the centre thereof, at equal distances from the upper surface of the flange and lower surface of the head, and $3\frac{1}{8}$ inches from the end of the rail to the centre of the first hole, and of 5 inches from the centre of the first hole to the centre of the second hole.

"9. That the lengths of rails at 60° Fahrenheit shall be kept within $\frac{1}{8}$ of an inch of the standard lengths, which are 30 feet, $27\frac{1}{2}$ feet, and 25 feet. That not more than 10 per cent. of rails of the shorter lengths shall be delivered in any one contract.

"10. That the rough edges produced at the ends of the rails by the saw shall be well trimmed off and filed.

"11. That rails are so straightened that they will insure a perfectly straight track.

"12. In order to get the exact length, as mentioned in paragraph No. 9, it is necessary that the rails should be cut off at as near a uniform temperature as practicable, that the blooms should not be detained in the process of rolling, and should go to the saws at a uniform temperature.

"The causes for a *temporary* rejection of the rails are:

"1. Bad straightening.

"2. Imperfect ends (which, after being cut off, would give a perfect rail of one of the standard short lengths).

"3. Missing test reports.

"4. A variation of more than $\frac{1}{8}$ of an inch from the standard lengths.

"The causes for a *permanent* rejection of a rail are:

"1. A bad test report, showing a deficiency or excess of carbon.

"2. The presence of a flaw of $\frac{1}{4}$ of an inch in depth in any part of the rail.

"3. The presence of such other imperfection as may involve a possibility of the rail breaking in the track.

"4. A greater variation between the rail and splice-bars than is allowed in paragraph No. 6."

Dr. Charles B. Dudley, of the Pennsylvania Railroad, found from analyses of the best steel rails that the following are the best proportions of the elements:

Limit of phosphorus, $\frac{1}{10}$ per cent.
" silicon, $\frac{4}{100}$ per cent.
" carbon, $\frac{25}{100}$ to $\frac{35}{100}$ per cent.
" manganese, $\frac{30}{100}$ to $\frac{40}{100}$ per cent.

The average of good rails was: Carbon, .287; phosphorus, .077; manganese, .369, and silicon, .044. The tensile strength should be above 65,000 pounds per square inch, and the elongation before breaking about 30 per cent., the ductility, as indicated by this last quality, being specially important. These proportions of the elements, however, are not accepted by other engineers as being exact; there may be more phosphorus if the amount of manganese is also increased. Sulphur and copper have not been stated, because if there is enough of either to do harm the steel will be "hot short," and cannot be rolled. (See discussion in Transactions American Society of Mining Engineers, Vols. VII. and VIII.)

Dr. Dudley likewise recommends that for a mechanical test, pieces be cut from the web of the rail 12 inches long, 1½ inch wide and ½ inch thick, and supported on "knife edges" 10 inches apart, and bent by a knife edge at equal distances from the supports; they should stand not over 3,000 pounds, and bend not less than 130° without rupture.

The tests applied to rails in Germany are as follows: They must sustain 20 tons without permanent flexure, when resting on supports 1 metre apart; they must bear two strokes of 1,100 pounds falling 13 feet without breaking, and one stroke of the same weight falling 5 feet without injury; they must also bend cold 2 inches without cracks, and curve $\frac{9}{10}$ of an inch in 9 feet 10 inches.

WATER STATIONS.

Passenger engines on the Middle Division of the Pennsylvania Railroad, where the grades are very light, run at a rate of 35 miles per hour with seven cars; and, when making frequent stops, one tank of water, containing 2,400 gallons, lasts for two and a quarter hours, or 78 miles. The engines, however, take in water actually every 45 miles. A freight train on the same division, with a full tank, can run at a speed of $14\frac{1}{2}$ miles an hour for two hours fifty minutes, or $41\frac{1}{2}$ miles, with one tank of water. As, however, they have to stop at shorter intervals to allow passenger trains to pass, or to pass each other, they utilize the time of waiting in filling their tanks. On the Mountain Division of the same railroad, freight trains, with a full load on an almost continuous grade of $1\frac{8}{10}$ per cent., use a tankful of water, containing 3,000 gallons, in one hour fifteen minutes, or in going 15 miles. It is thus seen that 15 miles is the extreme distance apart for water stations with grades of 2 feet in 100, while 40 miles would do with very light grades. It would be well, however, if water is plenty, to have them every 10 miles, or oftener.

In a hilly country, streams can generally be dammed up, which will give a gravity supply. The outer slope of the dam may be built of stone, like a retaining wall, or may be of earth at its natural slope. The inner slope should be at the natural slope of clay in water, which is 3 to 1. There should be a layer of good clay on the inside, 2 feet thick. Cast-iron pipe, 6 inches in diameter, is used to convey the water from the dam to the track. This should have a fall of at least 1 or $1\frac{1}{2}$ feet per 100 for 1,000 feet from the dam. For the remaining distance, if the water is brought so far, the fall had better be not less than .333 per 100. If, however, it is impossible to give it a continuous down grade, it may be laid undulating, so long as no portion rises above the "hydraulic grade line," as defined in "Trautwine's Pocket Book." If it is so laid undulating, it will be necessary to place an air cock at every "summit," and a mud cock (blow-off cock) at every "valley." At the track, a "stand-pipe" or "plug" is placed, which rises to an outlet 9 feet above the rail. A valve controls the outlet, within reach of the engine driver. A piece of rubber hose, 7 inches in diameter, 10 feet long, is fastened on the end of the pipe, to insert into the opening in the tender.

It may often happen that a dam cannot be made, or there is not enough water in the stream to furnish a continuous supply. A tank is then placed at the side of the railroad. This is a tub made of white pine, 18 feet in diameter at the bottom and 17 at the top, 8 feet deep. The bottom is 3 inches thick and the staves $2\frac{1}{2}$ inches thick. There are 6 iron hoops, $\frac{1}{4} \times 3$ inches ; two placed close together at the base, and the others at intervals increasing toward the top. The bottom is let into a groove in the staves, but the ends of the staves are let into the floor, so that the bottom bears over its whole surface on the floor. The tub is supported on three trestles of 10×10 inches stuff, placed 6 feet 6 inches apart, on walls 18 inches thick, built parallel to the track, and finished off 1 foot above the rail. On these trestles, joists 4×12 inches are placed 1 foot apart, which support the floor of 2-inch stuff, on which the bottom of the tub directly rests. Where a greater supply is required, or a more permanent structure, and an adjacent hill permits it, stone reservoirs are made, 40 feet in diameter and 8 feet deep. They are built below the surface of the ground. The walls are built of common mortar, with a lining of brick, well wet and thoroughly bedded in cement. The bottom is covered with a layer of stones about the size of a walnut, 4 inches deep, and made into a concrete with cement ; and when it is set another layer of the same thickness is put in. These reservoirs are covered with a house.

Where a gravity supply cannot be obtained, water must

be pumped into the reservoir with a steam engine or windmill.

Six-inch pipe, of a thickness of $\frac{7}{16}$ of an inch, is made to lie in lengths of 12 feet. Each joint requires 8 pounds of lead and $\frac{1}{4}$ of a pound of "gasket," or loosely twisted rope, which comes for the purpose.

COALING STATIONS.

A freight engine with its load, on very light grades, consumes about 160 bushels of bituminous coal in going 131 miles, and tenders hold about 80 bushels, or 8,000 pounds.

This will give some basis for calculating the distances at which coaling stations must be provided.

PASSENGER STATIONS.

Description of Cresson Passenger Station, Pennsylvania Railroad:

One story high, 70 × 40 × 12 feet high, with sloping roof. Posts or "studs" are set 6 × 7 inches at the corners, and 5 × 6 inches at points $5\frac{1}{2}$ feet apart. These are braced by horizontal pieces, 3 × 4 inches, placed about 4 feet apart, except at the windows and doors. Diagonal braces, 4 × 6 inches, are placed at the upper corners, framed into the posts and a beam 6 × 7 inches, which forms the tie-beam of the roof truss. The latter is a king post of 6 × 6 inch pieces, with secondary king-post trusses abutting toward the centre against a straining beam, all 4 × 6 inches. The ridge pole is 2 × 10 inches, with purlins 4 × 7, and rafters 3 × 5 inches, spaced 2 feet apart. Joists, 3 × 10, spaced $1\frac{1}{2}$ feet apart. Flooring, $1\frac{1}{4}$ inches, worked. Roof sheeting, 1 inch. Platform flooring, 2 inches. Platform joists, 3 × $9\frac{1}{2}$ inches. Weather boarding, $\frac{7}{8}$ of an inch thick, 9 inches wide, with $\frac{3}{8}$ of an inch stripping, 2 inches wide. Partition of $\frac{7}{8}$ of an inch stuff. Plastering lath, 3 feet long.

Water station at Gallitzin, Pennsylvania Railroad :

"Balloon frame," 22 feet 8 inches by 22 feet 8 inches by 18 feet 6 inches high. Wall plates, 3 × 8 inches. End posts, 4 × 4 inches. Studs, 2 × 4 inches, 18 inches apart. Diagonal pieces, 1½ × 3 inches, 2 feet 6 inches apart, measured vertically. Rafters, 2 × 6 inches. Ridge pole, 2 × 8 inches. Joists, 4 × 12 inches. Siding of ⅞ of an inch worked boards, tongued and grooved. Sheeting, ditto. Slate roof.

It may here be remarked that when a plank is nailed to a post or joist, or other wooden substance, a nail is used of such a length that it will go twice as far into the post or joist as the thickness of the plank. Thus, for ⅞-inch stuff use 2½ inches long or 8-penny nails, for 1-inch stuff use 3 inches long or 10-penny nails.

TELEGRAPH LINE.

The number of poles to the mile varies from 26 to 42. The size of wire varies from 320 to 380 pounds per mile. The poles should not be less than 5 inches in diameter at the top, nor less than 25 feet long. If green, they should be charred 5 feet from the bottom. If any are split at the lower end, the parts should be nailed together before putting in the ground ; otherwise, the spring of the wood will prevent the earth from packing around them. The cross-arms are made of 3 × 4 inches, 3 feet long, white pine, painted white, one bolt for each cross-arm, ½ inch diameter, 8 inches long, square head and nuts, and wrought washers.

Sixty miles of line will require at each end a battery of 15 cups (Grove's). These cups require to be re-charged twice a week. A battery of 30 cups requires one carboy or 200 pounds of nitric acid, 25 pounds of sulphuric acid, and 1 pound of zinc per cup, every month.

APPENDIX.

The following are the specifications from which the Columbia & Port Deposit Railroad was built:

GRADUATION.

1. Under this head will be included all excavations and embankments required for the formation of the road-bed; cutting all ditches or drains about or contiguous to the road; the foundations of culverts, and all small bridges or walls; the excavations and embankments necessary for reconstructing turnpikes or common roads, in cases where they are destroyed or interfered with in the formation of the railroad; and all other excavations or embankments connected with or incident to the construction of said railroad.

2. All cuttings shall be measured in the excavations and estimated by the cubic yard under the following heads, viz.: Earth, loose rock, solid rock.

Earth will include clay, sand, loam, gravel, and all other earthy matter, or earth containing loose stone or boulders intermixed which do not exceed in size 3 cubic feet.

Loose rock shall include all stone and detached rock, lying in separate and contiguous masses, containing not over 3 cubic yards; also all slate or other rock that can be quarried without blasting, although blasting may be occasionally resorted to.

Solid rock includes all rock occurring in masses exceeding 3 cubic yards, which cannot be removed without blasting.

3. The road will be graded for a single track, except where otherwise directed by the Engineer, with a road-bed of such width, and side slopes of such inclination as the Engineer shall in each case designate, and in conformity to such depths of cuttings and fillings as may have been or may hereafter be determined upon by said Engineer.

4. Earth, gravel and other materials taken from excavations (except when otherwise directed by the Engineer), shall be deposited in the adjacent embankment, the cost of removing which, when the haul is not more than 1,400 feet, will be included in the price paid for excavation; all material necessarily procured from without the road and deposited in the embankments will be paid for as embankment only, but all material necessarily procured from within the line of the railroad, and hauled more than 1,400 feet, will be paid for as excavation and also as an embankment. In procuring materials for embankment from without the line of the road, the place will be designated by the Engineer in charge of the work, and in excavating and removing it, care must be taken to injure or disfigure the land as little as possible. The embankment will be formed in layers of such depth, and material disposed and distributed in such manner as the Engineer may direct, with the required allowance for settling.

Materials necessarily wasted from the cuttings will be deposited in the vicinity of the road, according to the directions of the Engineer in charge, and if, during the progress of the work, the raft channel of the Susquehanna River should be obstructed, by blasting of rocks, sliding of earth, or from any other cause, the same shall be carefully removed, without delay by the contractor, at his own expense.

5. The ground to be occupied by the excavations and embankments, together with a space of 12 feet beyond the

slope stakes on each side, or 10 feet beyond the berm ditch, where one is required, will be cleared of all trees, brush and other perishable matter. Where the filling does not exceed 3 feet, the trees, stumps and saplings must be grubbed, but under all other portions of the embankment it will be sufficient that they be cut close to the earth. No separate allowance will be made for grubbing and clearing, but its cost will be included in the price for excavation and embankment.

6. Contractors, when directed by the Engineer in charge of the work, will deposit on the side of the road, or at such convenient points as may be designated, any stone or rock that they may excavate, and if in so doing they should deposit material required for embankment, the additional cost, if any, of procuring other materials from without the road will be allowed. All stone or rock excavated and deposited as above, together with all timber removed from the line of the road, will be considered the property of the Railroad Company, and the contractors upon the respective sections will be responsible for its safe keeping until removed by said Company, or until the work is finished.

7. The line of road or the gradients may be changed, if the Engineer shall consider such change necessary or expedient, and for any considerable alterations, the injury or advantage to the contractor will be estimated, and such allowance or deduction made in the prices as the Engineer may deem just and equitable; but no claim for an increase in prices of excavation or embankment on the part of the contractor will be allowed or considered, unless made in writing, before the work on that part of the section where the alteration has been made shall have been commenced. The Engineer may also, on the conditions last recited, increase or diminish the length of any section for the purpose

of more nearly equalizing or balancing the excavations and embankments.

8. Whenever the route of the railroad is traversed by public or private roads, commodious passing places must be kept open and in safe condition for use ; and in passing through farms the contractor must also keep up such temporary fences as will be necessary for the preservation of the crops.

MASONRY.

All masonry will be estimated and paid for by the perch of 25 cubic feet, and will be included under the following heads, viz.: Rectangular and arch culvert masonry, first and second quality bridge masonry, vertical and slope wall masonry and paving.

1. *Culvert Masonry.*—All rectangular culverts will be built dry (not less than $2\frac{1}{2} \times 3$ feet) as may be required by the Engineer ; the abutment will rest on a pavement of stone, set edgewise, of at least 10 inches in depth, confined and secured at the ends by deep curb-stones, which will be protected from undermining by broken stone, placed in such quantity and position as the Engineer may direct. The abutment walls will be not less than 2 feet thick, and built of good-sized, well-shaped stone, properly laid and bound together by stones, occasionally extending entirely through the walls. The upper course to have at least one-half of the stones headers ; and the stretchers in no case to be less than 12 inches wide ; no stone in this course to be less than 6 inches thick. The covering to be of sound, strong stone, at least 12 inches thick, and to lap not less than 10 inches on each abutment. The thickness of the covering stone and dimensions of the walls to be increased at the discretion of the Engineer, according to the height of the embankment and span of the culvert.

2. *Semi-Circular or other Culverts with Curved Arches.*—The foundations of these culverts, when the bottom of the pit is common earth, gravel, etc., will generally consist of a pavement formed of stone set edgewise, not less than 12 inches in depth, secured in the same manner as before described for rectangular culverts. When the foundation upon which a culvert is to be built is soft and compressible, and where it will at all times be covered with water, timber well hewed and from 8 to 12 inches in thickness (according to the span of the culvert) will be laid side by side crosswise upon longitudinal sills, and where a strong current will be forced through during floods, three courses of sheet piling are to be placed across the foundation, one course at each end and one in the middle, to be sunk from 3 to 6 feet below the top of the timber, according as the earth is more or less compact. The abutments are to be built of rubble work, the stones hammered on their beds and laid in courses; the stretchers in the face are to have beds of at least 15 inches, and they are to be not less than 2 feet long, measuring in the face of the wall; the headers will extend through the wall in cases where it does not exceed $3\frac{1}{2}$ feet thick, and they shall have not less than 18 inches length of face. There shall be not less than one header to every 7 feet of face, measured from centre to centre, and so arranged that a header in a superior course shall be placed between two headers in the course below; the backing stone shall be of large size and have parallel beds, laid so as to break joints with one another, and when the thickness of the wall exceeds three and a half feet, headers of the same dimensions as those in the face will be placed in the back of the wall in the proportion of one for every two headers in the face. The beds and joints of the arch stone are to be dressed so as to give an even bearing on each other, and to be laid in

courses throughout. The ring stone will be neatly cut and composed of alternate long and short bond stones of not less than three feet, and eighteen inches respectively. The parapet and wing walls will be built similarly to the abutments, and surmounted with a well-dressed coping r ot less than ten inches thick and three feet wide.

3. *Bridge Masonry.*—When rock foundation cannot be had for abutments and piers, the masonry shall be started upon hewn timber, sunken to such a depth as to protect it from decay and to prevent the possibility of underwashing. The timber platforms will be composed of one or more courses, according to the depth of the water, the height of the masonry, or other circumstances of which the Engineer shall judge and determine. The masonry will be of two qualities, either to be adopted at the *discretion* of the Engineer. First quality will be rock range work. The stone to be accurately squared, jointed and bedded, and laid in courses of not less than twelve inches thick, nor exceeding twenty inches in thickness, regularly decreasing from bottom to top of pier or abutment. The stretchers shall in no case have less than sixteen inches bed for a twelve-inch course, and for all courses above sixteen inches, at least as much bed as face; they shall generally be at least four feet in length. The headers will be of similar size with the stretchers, and shall hold the size in the heart of the wall, that they show on the face, and be so arranged as to occupy one-fifth of the face of the wall, and they will be similarly disposed in the back. When the thickness of the wall will admit of their interlocking they will be disposed in that manner. When the wall is too thick to admit of that arrangement, stones not less than four feet in length will be placed transversly in the heart of the wall to connect the two opposite sides of it. The stones for the heart of the wall will be of the

same thickness as those in the face and back, and must be well fitted to their places; any remaining interstices will be filled with small stones or chips. The face stones will, with the exception of the draught, be generally left with the face as they come from the quarry, unless the projections above the draught should exceed two inches, in which case they shall be roughly scabbled down to that point. The abutments or piers, and such portions of them as the Engineer may direct, shall be covered with a course of coping not less than twelve inches thick, well dressed, and fastened together with clamps of iron.

The second quality of bridge masonry will be rubble work, and will consist of stones containing generally six cubic feet each, so disposed as to make a firm and compact work, and no stone in the work shall contain less than two cubic feet except for filling up the interstices between the large blocks in the heart of the wall; at least one-fifth of the face shall be composed of headers extending full size four feet into the wall, and from the back the same proportion and of the same dimensions, so arranged that a header in the back shall be between two headers in the face. The corner stones shall be neatly hammer-dressed, so as to have horizontal beds and vertical joints.

4. *Vertical and Slope Wall.*—The vertical walls will be good dry rubble work, of such dimensions, and built with such batter, as the Engineer may direct. Slope walls will be built of such thickness and slope as may be required by the Engineer. No stones, however, to be used in its construction which do not reach through the wall, nor any less than six inches in thickness by twelve inches long; the bed of the stone to be placed at right angles with the face of the bank, the joints close and free from spalls.

5. In all masonry the stones must be of a hard and durable quality, of good size and shape to be approved of by

the Division or Principal Assistant Engineer. Such portions of the masonry as the Engineer may require to be laid in lime mortar or hydraulic cement will be so laid, the Railroad Company furnishing or paying for the lime and cement used. If in the progress of the masonry, an increase in the number of headers specified should be required by the Engineer, such additional number shall be laid in the work as he shall designate.

6. The price per perch for masonry shall in every case include the furnishing of all materials (except lime and cement); the cost of scaffolding, centering, etc., and all expenses attending the delivering of these materials and all risks from floods or otherwise.

7. *Rip Rap.*—All rock in excavation to be deposited upon the river side of embankments, within the distance considered an overhaul, of such depth, and at such places as the Engineer may direct, without any extra compensation being allowed, excepting whenever the party of the first part may, or shall be required by the party of the second part, or the said Engineer, to use the material out of the rock cuts, for rip rapping any embankments that may have been made of earth, he shall be allowed a price per cubic yard for such rip rap, in addition to the rock price. If an overhaul, it shall be paid for as excavation and embankment. In classification, it shall be known as "rip rap from excavation."

Where the excavation of the road bed does not furnish sufficient stone for the protection of walls and embankments, the same shall be procured, at such places and disposed in such manner as the Engineer may direct, and a price per cubic yard paid therefor.

8. *Ballast.*—The ballast must be of good hard stone, to be approved by the Engineer. It must be well broken into cubical pieces of such size as to pass through a ring of

three inches in diameter. It must be placed on the road bed, of such width and depth as the Engineer may direct.

9. The quantities exhibited to the contractor at the letting are, from the necessity of the case, *merely approximate;* they furnish only general information, and will in no way govern or affect the final estimate of the work, which will be made out on its completion, from actual measurements and established facts, not now in the possession of any one, nor possible to be obtained at the time of drawing up this specification.

10. No charge shall be made by the contractor for hindrances or delay, from any cause, in the progress of any portion of the work in this contract, but it may entitle him to an extension of time allowed for completing this work sufficient to compensate for the detention, to be determined by the Chief Engineer, provided he shall give the Engineer in charge immediate notice in writing of the cause of the detention.

Nor shall any claim be allowed for extra work unless the same shall be done in pursuance of a written order from the Engineer in charge, and the claim made at the first settlement after the work was executed, unless the Chief Engineer at his discretion should direct the claim or such part as he may deem just and equitable to be allowed.

11. Any work which the Engineer in charge may require done, under this contract, and for which there is no specific price named herein, shall be paid for on the estimate, and at a value fixed by said Engineer, subject to the approval of the Chief Engineer.

ON THE MOST ECONOMICAL HEIGHT OF BRIDGE TRUSS.

(*From Proceedings of the Engineers' Club, of Philadelphia.*)

In many cases of bridge design, the height of the truss is fixed by some extraneous local condition, which does not belong to the bridge itself; such as, the available head room, either between the trusses or under them, or a desire to limit the area exposed to the wind, etc. In the great majority of cases, however, the engineer has a great latitude in deciding on the ratio of the height of the truss to the span. This ratio is generally chosen by copying some existing bridge, and an arbitrary one-eighth or one-tenth the span is chosen, without considering very thoroughly what the proper proportion should be. For a given span, a given style of truss, a given number of panels, and a given load, there is a certain height of truss which will require least material, and is therefore the cheapest. To discover what this particular height should be, is not a very difficult problem. The ratio of height to the length of a panel is expressed by a letter m, and the stresses on the various parts written, being functions of m. The proper sections being found from the stresses, are multiplied by the lengths of the various pieces, and then by the price per cubic unit of the material, and added together. The value of m corresponding to the least value of this function, is found by the principle of minima in the calculus, making the first differential coefficient equal to zero, etc.

To apply this to the Howe truss,

Let $n =$ the number of panels.

$l =$ the length of a panel.

$W =$ the weight on one panel due to the uniform load.

$W' =$ the weight on one panel due to the variable load.

$h =$ the least dimension of a brace.

$h' =$ the least dimension of the upper chord.

$p =$ the cost in dollars per 1,000 ft., board measure, of lumber.

$p' =$ the cost in cents, per pound, of the iron.

$m =$ the height of truss divided by the length of one panel.

1,000 lbs. = the safe load on the wood.

10,000 lbs. = the safe load on the iron, and let Gordon's formula be used for calculating the strengths of the chord and braces, each being proportioned, in each panel, to its safe load.

We then have for the total cost of one-half truss:

$$\left[\begin{array}{l}\dfrac{1}{1000}\left\{\begin{array}{l}(n-1)(W+W')+(n-3)W+\\ \dfrac{(n-2)(n-1)}{n}W'+(n-5)W\\ +\dfrac{(n-3)(n-2)}{n}W'+(n-7)\\ W+\dfrac{(n-4)(n-3)}{n}W'+\&\text{c.,}\\ \text{until the algebraic sum of}\\ \text{these last two terms is a mi-}\\ \text{nus quantity.}\end{array}\right\}l\left(1+\dfrac{l^2}{250\,h^2}\right)\\ +\dfrac{1}{1000}\left\{\begin{array}{l}[n-1]+[(n-1)+\\ (n-3)]+[(n-1)+\\ (n-3)+(n-5)]+\\ \&\text{c., until the last}\\ \text{term in the brack-}\\ \text{ets }[\ldots]\text{ is 3.}\end{array}\right\}(W+W')\,l\left(1+\dfrac{l^2}{250\,h'^2}\right)\\ +\dfrac{1}{2000}\left\{\begin{array}{l}[n-1]+[(n-1)+(n-3)]+\\ [(n-1)+(n-3)+(n-5)]+\\ \&\text{c.,}\\ \text{until the last term in the}\\ \text{brackets }[\ldots]\text{ is 1.}\end{array}\right\}(W+W')\,l\end{array}\right]\times\dfrac{p}{1440}\times\dfrac{1}{m}$$

$$+\left[\begin{array}{l}+\dfrac{1}{1000}\left\{\begin{array}{l}(n-1)(W+W')+(n-3)W\\ +\dfrac{(n-2)(n-1)}{n}W''+(n-5)W+\\ \dfrac{(n-3)(n-2)}{n}W''+\&\text{c.,}\\ \text{until the algebraic sum of these}\\ \text{last two terms is a minus quantity.}\end{array}\right\}l\left(1+\dfrac{2\,l^2}{250\,h^2}\right)\dfrac{p}{1440}\\ +\dfrac{1}{20000}\left\{\begin{array}{l}(n-1)(W+W')+(n-3)W+\dfrac{(n-2)(n-1)}{n}\\ W''+(n-5)\,W+\dfrac{(n-3)(n-2)}{n}W''+\\ \&\text{c.,}\\ \text{until the last term in the numerator}\\ \text{is 3.}\end{array}\right\}l\dfrac{28}{100}\,p'\end{array}\right]\times m$$

$$+ \frac{1}{1000} \left\{ \begin{array}{l} (n-1)(W+W'')+(n-3)W'+ \\ \frac{(n-2)(n-1)}{n}W'+(n-5)W'+ \\ \frac{(n-3)(n-2)}{n}W'+ \&c., \\ \text{until the algebraic sum of these last} \\ \text{two terms is a minus quantity.} \end{array} \right\} l \frac{l^2}{250 h^2} \frac{p}{1440} \times m^3$$

This may be written in the form:

$$v = \frac{a}{m} + bm + cm^3$$

Putting its first differential coefficient equal to zero, we obtain the value of

$$m = \sqrt{\frac{\sqrt{12\,ac + b^2} - b}{6\,c}}$$

We can now apply these equations to some practical examples.

The variable load may be taken, for a railroad bridge at $1\frac{1}{2}$ tons per lineal foot, which gives in pounds $2\,W'' = 3360 \frac{l}{12}$ or $W' = 140\,l$.

By using the empirical formula of a previous page for calculating W, reducing to pounds and inches.

$$W = \frac{280}{3} \left\{ \frac{1 + \frac{nl}{156}}{17 + \sqrt{\frac{12}{nl}}} \right\} l$$

If $n = 12, l = 120, h = 8$ and $h' = 11, p = 27, p' = 5$, we have $W = 6,704$, $W'' = 16,800$, $m = .96$ and cost of one-half truss = \$342.89. If $n = 6, l = 240$, $W = 13,408$ and $W'' = 33,600$, the other quantities being as before, we have $m = .72$, and cost of one-half truss = \$501.78. If, in the last case, however, we make $h = 12$ and $h' = 20$, we have $m = .73$ and cost of one-half truss = \$310.33, which shows the result of using pieces of large diameter in the compression parts. It would seem, then, that when only small scantlings can be obtained, it would be better to make a large number of panels, and when large scantlings are procurable, to increase the length of the panels. It should be observed, however, that the load is supposed to be applied to the truss only at the panel points. As a matter of fact, it is usually applied on a chord, and the additional material required in the chord on this account would be proportionately greater in a long panel than in a shorter one. It is better, however, to use separate pieces to support the load between the panel points, and so confine the chords to their legitimate duty of resisting the stresses in the direction of their lengths due to the structure acting as a truss only.

www.ingramcontent.com/pod-product-compliance
Lightning Source LLC
Chambersburg PA
CBHW030335170426
43202CB00010B/1134